Tropical Nature
and Other Essays

ALFRED RUSSEL WALLACE

CAMBRIDGE
UNIVERSITY PRESS

CAMBRIDGE UNIVERSITY PRESS

Cambridge, New York, Melbourne, Madrid, Cape Town,
Singapore, São Paolo, Delhi, Mexico City

Published in the United States of America by Cambridge University Press, New York

www.cambridge.org
Information on this title: www.cambridge.org/9781108053136

© in this compilation Cambridge University Press 2013

This edition first published 1878
This digitally printed version 2013

ISBN 978-1-108-05313-6 Paperback

CAMBRIDGE LIBRARY COLLECTION

Books of enduring scholarly value

Life Sciences

Until the nineteenth century, the various subjects now known as the life
sciences were regarded either as arcane studies which had little impact
on ordinary daily life, or as a genteel hobby for the leisured classes. The
increasing academic rigour and systematisation brought to the study of
botany, zoology and other disciplines, and their adoption in university
curricula, are reflected in the books reissued in this series.

Tropical Nature and Other Essays

Sometimes referred to as 'the grand old man of science', Alfred Russel Wallace
(1823–1913) was a naturalist, evolutionary theorist, and friend of Charles
Darwin. In this study of tropical flora and fauna, he takes the reader on
a tour of the equatorial forest belt – the almost continuous band of forest
that stretches around the world between the tropics. There, chameleon-like
caterpillars alter the colours of their cocoons, parasitical trees override their
hosts with spectacular aerial root systems, and some of the most pressing
questions of Victorian evolutionary science arise: how do animals and
plants come to be brightly coloured? Can their adaptations provide clues
about past geological eras? And was Darwin wholly correct in his theory of
sexual selection? First published in 1878, Wallace's book is a skilfully written
reflection of contemporary naturalism, still highly readable and relevant to
students in the history of science.

Cambridge University Press has long been a pioneer in the reissuing of out-of-print titles from its own backlist, producing digital reprints of books that are still sought after by scholars and students but could not be reprinted economically using traditional technology. The Cambridge Library Collection extends this activity to a wider range of books which are still of importance to researchers and professionals, either for the source material they contain, or as landmarks in the history of their academic discipline.

Drawing from the world-renowned collections in the Cambridge University Library and other partner libraries, and guided by the advice of experts in each subject area, Cambridge University Press is using state-of-the-art scanning machines in its own Printing House to capture the content of each book selected for inclusion. The files are processed to give a consistently clear, crisp image, and the books finished to the high quality standard for which the Press is recognised around the world. The latest print-on-demand technology ensures that the books will remain available indefinitely, and that orders for single or multiple copies can quickly be supplied.

The Cambridge Library Collection brings back to life books of enduring scholarly value (including out-of-copyright works originally issued by other publishers) across a wide range of disciplines in the humanities and social sciences and in science and technology.

Selected titles by Darwin and other participants
in the early debates on evolutionary theory, available in the
CAMBRIDGE LIBRARY COLLECTION

Candolle, Augustin Pyramus de, and Sprengel, Kurt: *Elements of the Philosophy of Plants* (1821) [ISBN 9781108037464]

Cuvier, Georges, translated by Robert Kerr: *Essay on the Theory of the Earth* (1815) [ISBN 9781108005555]

Darwin, Charles: *Geological Observations on South America* (1846) [ISBN 9781108027144]

Darwin, Charles: *Geological Observations on the Volcanic Islands, Visited During the Voyage of H.M.S. Beagle* (1844) [ISBN 9781108072335]

Darwin, Charles: *Insectivorous Plants* (1875) [ISBN 9781108004848]

Darwin, Charles: *Journal of Researches into the Natural History and Geology of the Countries Visited during the Voyage of H.M.S. Beagle* (second edition, 1845) [ISBN 9781108038065]

Darwin, Charles: *Monographs on the Fossil Lepadidae, Balanidae and Verrucidae* (1851) [ISBN 9781108004824]

Darwin, Charles: *On the Movements and Habits of Climbing Plants* (1865) [ISBN 9781108003599]

Darwin, Charles: *On the Various Contrivances by which British and Foreign Orchids are Fertilised by Insects* (1862) [ISBN 9781108027151]

Darwin, Charles: *The Different Forms of Flowers on Plants of the Same Species* (1877) [ISBN 9781108018272]

Darwin, Charles: *The Effects of Cross and Self Fertilisation in the Vegetable Kingdom* (1876) [ISBN 9781108005258]

Darwin, Charles, edited by Francis Darwin: *The Expression of the Emotions in Man and Animals* (second edition, 1890) [ISBN 9781108004831]

Darwin, Charles: *The Formation of Vegetable Mould through the Action of Worms* (1881) [ISBN 9781108005128]

Darwin, Charles, edited by Francis Darwin: *The Life and Letters of Charles Darwin* (3 vols., 1887) [ISBN 9781108003421]

Darwin, Charles: *The Origin of Species* (sixth edition, 1872) [ISBN 9781108005487]

Darwin, Charles: *The Structure and Distribution of Coral Reefs* (1842)
[ISBN 9781108065627]

Darwin, Charles: *The Variation of Animals and Plants under Domestication*
(2 vols., 1868) [ISBN 9781108014243]

Darwin, Charles: *The Descent of Man and Selection in Relation to Sex* (2 vols., 1871)
[ISBN 9781108005111]

Darwin, Charles, edited by Francis Darwin: *The Foundation of the Origin of Species*
(1909) [ISBN 9781108004886]

Darwin, Charles, edited by Francis Darwin: *The Power of Movement in Plants* (1880)
[ISBN 9781108003605]

Darwin, Charles, Henslow, John Stevens, and Sedgwick, Adam:
The Teaching of Science at Cambridge (1846) [ISBN 9781108002004]

Geikie, Archibald: *Charles Darwin as Geologist* (1909) [ISBN 9781108002578]

Lamarck, Jean-Baptiste Pierre Antoine de Monet de: *Philosophie zoologique*
(2 vols., 1809) [ISBN 9781108038041]

Lyell, Charles: *Principles of Geology* (3 vols., 1830–3) [ISBN 9781108001342]

Lyell, Charles: *The Geological Evidences of the Antiquity of Man* (1863)
[ISBN 9781108003971]

Romanes, George John: *Mental Evolution in Animals* (1883) [ISBN 9781108037877]

Romanes, George John: *Mental Evolution in Man* (1888) [ISBN 9781108037976]

Romanes, George John: *Darwin, and after Darwin* (3 vols., 1893–7)
[ISBN 9781108038126]

Romanes, George John, edited by E.D. Romanes: *The Life and Letters of George John Romanes* (1896) [ISBN 9781108037891]

Wallace, Alfred Russel: *Contributions to the Theory of Natural Selection* (1870)
[ISBN 9781108001540]

Wallace, Alfred Russel: *Darwinism* (1889) [ISBN 9781108001328]

For a complete list of titles in the Cambridge Library Collection please visit:
http://www.cambridge.org/features/CambridgeLibraryCollection/books.htm

TROPICAL NATURE,

AND OTHER ESSAYS.

TROPICAL NATURE,

AND

OTHER ESSAYS.

BY

ALFRED R. WALLACE.

AUTHOR OF "THE MALAY ARCHIPELAGO," "THE GEOGRAPHICAL DISTRIBUTION OF ANIMALS,"
"CONTRIBUTIONS TO THE THEORY OF NATURAL SELECTION," ETC., ETC.

London:
MACMILLAN AND CO.
1878.

LONDON :
R. CLAY, SONS, AND TAYLOR,
BREAD STREET HILL.

THE TROPICS.

Land of the Sun ! where joyous green-robed Spring
And leaf-crowned Summer deck the Earth for ever ;
No Winter stern their sweet embrace to sever
And numb to silence every living thing,
But bird and insect ever on the wing,
Flitting 'mid forest glades and tangled bowers,
While the life-giving orb's effulgent beams
Through all the circling year call forth the flowers.
Here graceful palms, here luscious fruits have birth ;
The fragrant coffee, life-sustaining rice,
Sweet canes, and wondrous gums, and odorous spice ;
While Flora's choicest treasures crowd the teeming earth.
Beside each cot the golden Orange stands,
And broad-leaved Plantain, pride of Tropic lands.

ENGLAND.

Sweet changing Seasons ! Winter cold and stern,
Fair Spring with budding leaf and opening flower,
And Summer when the sun's creative power
Brings leafy groves and glades of feathery fern,
The glorious blossoms of sweet-scented May,
The flowery hedgerows and the fragrant hay,
And the wide landscape's many-tinted sheen.
Then Autumn's yellow woods and days serene ;
And when we've gathered in the harvest's treasure,
The long nights bring us round the blazing hearth,
The chosen haunt of every social pleasure.
Land of green fields and flowers ! Thou givest birth
To shifting scenes of beauty, which outshine
Th' unvarying splendours of the Tropic's clime.

PREFACE.

THE luxuriance and beauty of Tropical Nature is a well-worn theme, and there is little new to say about it. The traveller and the naturalist have combined to praise, and not unfrequently to exaggerate the charms of tropical life—its heat and light, its superb vegetable forms, its brilliant tints of flower and bird and insect. Each strange and beautiful object has been described in detail; and both the scenery and the natural phenomena of the tropics have been depicted by master hands and with glowing colours. But, so far as I am aware, no one has yet attempted to give a general view of the phenomena which are *essentially* tropical, or to determine the causes and conditions of those phenomena. The local has not been separated from the general, the accidental from the essential; and, as a natural result, many erroneous ideas have become current as to what are really the characteristics of the tropical as distinguished from the temperate zones.

In the present volume I have attempted to supply this want; and for my materials have drawn chiefly on my own twelve years' experience of the eastern and western tropics of the equatorial zone, where the characteristic phenomena of tropical life are fully manifested.

So many of the most remarkable forms of life are now restricted to the tropics, and the relations of these to extinct types which once inhabited the temperate zones open up so many interesting questions as to the past history of the earth, that the present inquiry may be considered a necessary preliminary to a study of the problem—how to determine the climates of geologic periods from the character of their organic remains. This part of the subject is however both complex and difficult, and I have only attempted to indicate what seem to me the special physical conditions to which the existing peculiarities of tropical life are mainly due.

The three opening chapters treat the subject under the headings of climate, vegetation, and animal life. The conditions and causes of the equatorial climate are discussed in some detail, and the somewhat complex principles on which it depends are popularly explained. In the chapters on plant and animal life, the general aspects and relations of their several component elements have been dwelt upon; all botanical and zoological details and nomenclature being excluded, except so far as was absolutely necessary to give precision to the descriptions and to enable us to deduce from them some conclusions of importance.

The remaining chapters have all a more or less direct connection with the leading subject. The family of humming-birds is taken as an illustration of the luxuriant development of allied forms in the tropics, and as showing the special mode in which natural selection has acted to bring about considerable changes in a limited period. The discussion on the nature and origin of the colours of animals and plants, is intended to show how far and in what way these are dependent on the climate and physical conditions of the tropics. The chapter entitled "By-paths in the Domain of Biology" contains an account of certain curious relations of colour to locality, which are almost exclusively manifested within the tropical zones ; while the essay on "Distribution of Animals and Geographical Changes," elucidates the relations of the several continents in past time, and the probable origin of many of the groups now characteristic of tropical or of temperate regions.

While discussing the general laws and phenomena of colour in the organic world, and its special developments among certain groups of animals, I have been led to a theory of the diverse colours of the sexes and of the special ornaments and brilliant hues which dis-

tinguish certain male birds and insects, which is directly
opposed to the view held by Mr. Darwin and so well
explained and illustrated in his great work on " The
Descent of Man and on Selection in Relation to Sex."
Being strongly impressed with the importance and
fundamental truth of this theory, I published my first
sketch of the subject in *Macmillan's Magazine* in order
that it might have the benefit of criticism before making
it public in a more permanent form.   Taking advantage
of some suggestions from Mr. Darwin and from a
few other correspondents, I have made considerable
additions to the original essay and have rearranged,
and I trust strengthened the argument, which I now
hope may attract the attention of all who are interested
in the subject.   I may be allowed here to remark, that
my theory cannot be properly understood without
reading the whole chapter on " The Colours of
Animals ; " because the view set forth and illustrated
in the first part of that chapter—that colour in nature
is normal, and that its presence hardly requires to
be accounted for so much as its absence—is an essential
part of the theory.

CROYDON, *April*, 1878.

CONTENTS.

VII. BY-PATHS IN THE DOMAIN OF BIOLOGY.

VIII. THE DISTRIBUTION OF ANIMALS AS INDICATING GEOGRAPHICAL CHANGES.

TROPICAL NATURE,

AND OTHER ESSAYS.

TROPICAL NATURE,

AND OTHER ESSAYS.

I.

THE CLIMATE AND PHYSICAL ASPECTS OF THE EQUATORIAL ZONE.

The three Climatal Zones of the Earth—Temperature of the Equatorial Zone
—Causes of the Uniform High Temperature near the Equator—Influence
of the Heat of the Soil—Influence of the Aqueous Vapour of the Atmosphere—Influence of Winds on the Temperature of the Equator—Heat due
to the Condensation of Atmospheric Vapour—General features of the
Equatorial Climate—Uniformity of the Equatorial Climate in all parts of
the globe—Effects of Vegetation on Climate—Short Twilight of the Equatorial Zone—The aspect of the Equatorial Heavens—Intensity of meteorological phenomena at the Equator—Concluding Remarks.

IT is difficult for an inhabitant of our temperate land
to realize either the sudden and violent contrasts of the
arctic seasons or the wonderful uniformity of the equatorial climate. The lengthening or the shortening days,
the ever-changing tints of spring, summer, and autumn,
succeeded by the leafless boughs of winter, are constantly
recurring phenomena which represent to us the established course of nature. At the equator none of
these changes occur; there is a perpetual equinox and

a perpetual summer, and were it not for variations in the quantity of rain, in the direction and strength of the winds, and in the amount of sunshine, accompanied by corresponding slight changes in the development of vegetable and animal life, the monotony of nature would be extreme.

In the present chapter it is proposed to describe the chief peculiarities which distinguish the equatorial from the temperate climate, and to explain the causes of the difference between them,—causes which are by no means of so simple a nature as are usually imagined.

The three great divisions of the earth—the tropical, the temperate, and the frigid zones, may be briefly defined as the regions of uniform, of variable, and of extreme physical conditions respectively. They are primarily determined by the circumstance of the earth's axis not being perpendicular to the plane in which it moves round the sun ; whence it follows that during one half of its revolution the north pole, and during the other half the south pole, is turned at a considerable angle towards the source of light and heat. This inclination of the axis on which the earth rotates is usually defined by the inclination of the equator to the plane of the orbit, termed the obliquity of the ecliptic. The amount of this obliquity is $23\frac{1}{2}$ degrees, and this measures the extent on each side of the equator of what are called the tropics, because within these limits the sun becomes vertical at noon twice a year, and at the extreme limit once a year, while beyond this distance it is never vertical. It will be evident, however, from the nature of the case, that the two lines which mark the limits of the geographical " tropics " will not define any abrupt change of climate or physical conditions, such as characterise the tropical and temperate zones in their full development. There will be a gradual transition from one to the other, and in order to study them separately and contrast their special features we must only take into account the portion of each in which these are most fully exhibited. For the temperate zone we may take all countries situated between 35° and 60° of latitude, which in Europe will include every place between Christiana and Algiers, the districts further south forming a transitional belt in which temperate and tropical features are combined. In order to study the special features of tropical nature, on the other hand, it will be advisable to confine our attention mainly to that portion of the globe which extends for about twelve degrees on each side of the equator, in which all the chief tropical phenomena dependent on astronomical causes are most fully manifested, and which we may distinguish as the " equatorial zone." In the debateable ground between these two well contrasted belts local causes have a preponderating influence ; and it would not be difficult to point out localities within the temperate zone of our maps, which exhibit all the chief characteristics of tropical nature to a greater degree than other localities which are, as regards geographical position, tropical.

Temperature of the Equatorial Zone.—The most characteristic, as it is the most important feature in the physical conditions of the great equatorial zone is the wonderful uniformity of its temperature, alike throughout the changes of day and night, and from one part of the year to another. As a general rule, the greatest heat of the day does not exceed 90° or 91°

Fahr., while it seldom falls during the night below 74° Fahr. It has been found by hourly observations carried on for three years at the meteorological observatory established by the Dutch government at Batavia, that the extreme range of temperature in that period was only 27° Fahr., the maximum being 95° and the minimum 68°. But this is, of course, very much beyond the usual daily range of the thermometer, which is, on the average, only a little more than 11° Fahr.; being 12·6° in September when it is greatest, and only 8·1° in January, when it is least.

Batavia, being situated between six and seven degrees south of the equator, may be taken as affording a fair example of the climate of the equatorial zone; though, being in an island, it is somewhat less extreme than many continental localities. Observations made at Para, which is continental and close to the equator, agree however very closely with those at Batavia; but at the latter place all the observations were made with extreme care and with the best instruments, and are therefore preferred as being thoroughly trustworthy.[1] The accompanying diagram, showing by curves the monthly means of the highest and lowest daily temperatures at Batavia and London, is very instructive; more especially when we consider that the maximum of temperature is by no means remarkably different in the two places, 90° Fahr. being sometimes reached with us and not being often very much exceeded at Batavia.

[1] "Observations Made at the Magnetical and Meteorological Observa at Batavia. Published by order of the Government of Netherlands I: Vol. I. Meteorological, from Jan. 1866 to Dec. 1868; and Magnetical, July 1867 to June 1870. By Dr. P. A. Bergsma. Batavia, 1871." fine work is entirely in English.

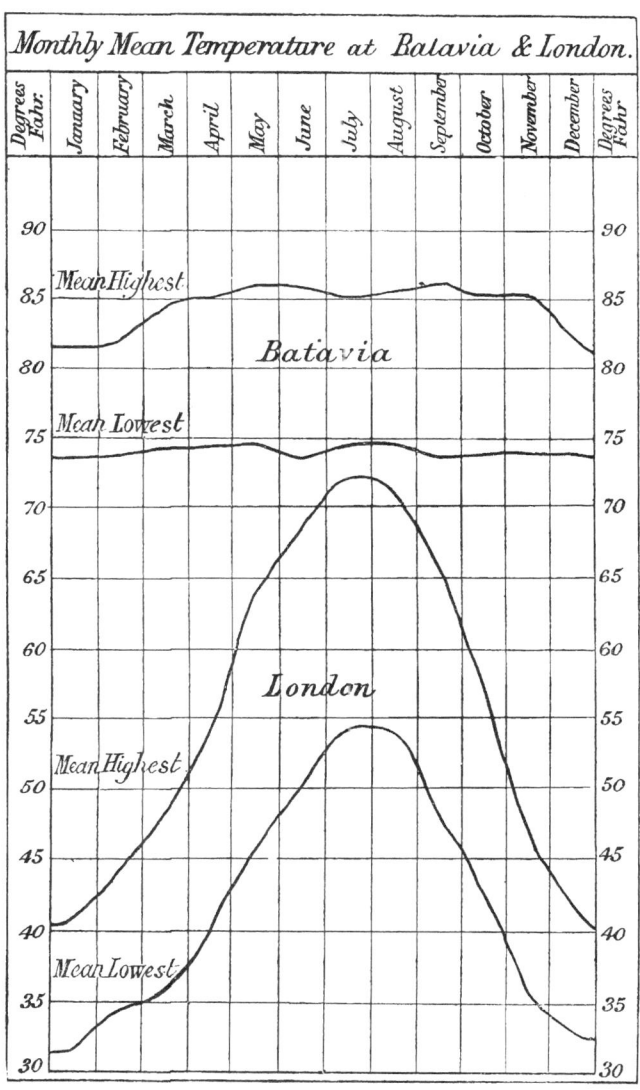

Monthly Mean Temperature at Batavia & London.

Causes of the Uniform High Temperature near the Equator.—It is popularly supposed that the uniform high temperature of the tropics is sufficiently explained by the greater altitude, and therefore greater heating-power, of the midday sun ; but a little consideration will show that this alone by no means accounts for the phenomenon. The island of Java is situated in from six and a half to eight and a half degrees of south latitude, and in the month of June the sun's altitude at noon will not be more than from 58° to 60°. In the same month at London, which is fifty-two and a half degrees of north latitude, the sun's noonday altitude is 62°. But besides this difference of altitude in favour of London there is a still more important difference ; for in Java the day is only about eleven and a half hours long in the month of June, while at London it is sixteen hours long, so that the total amount of sun-heat received by the earth must be then very much greater at London than at Batavia. Yet at the former place the mean temperature of the day and night is under 60° Fahr., while in the latter place it is 80° Fahr., the daily maximum being on the average in the one case about 68° and in the other about 89°.

Neither does the temperature at the same place depend upon the height of the sun at noon ; for at Batavia it is nearly vertical during October and February, but these are far from being the hottest months, which are May, June, and September ; while December, January, and February are the coldest months, although then the sun attains nearly its greatest altitude. It is evident, therefore, that a difference of 30° in the altitude of the sun at noon has no apparent influence in raising the temperature of a place near the equator, and we must therefore conclude that other agencies are at work which often completely neutralise the effect which increased altitude must undoubtedly exert.

There is another important difference between the temperate and tropical zones, in the direct heating effect of the sun's rays independently of altitude. In England the noonday sun in the month of June rarely inconveniences us or produces any burning of the skin ; while in the tropics, at almost any hour of the day, and when the sun has an elevation of only 40° or 50°, exposure to to it for a few minutes will scorch a European so that the skin turns red, becomes painful, and often blisters or peels off. Almost every visitor to the tropics suffers from incautious exposure of the neck, the leg, or some other part of the body to the sun's rays, which there possess a power as new, as it is at first sight inexplicable, for it is not accompanied by any extraordinary increase in the temperature of the air.

These very different effects, produced by the same amount of sun-heat poured upon the earth in different latitudes is due to a combination of causes. The most important of these are, probably,—the constant high temperature of the soil and of the surface-waters of the ocean,—the great amount of aqueous vapour in the atmosphere,—the great extent of the intertropical regions which cause the winds that reach the equatorial zone to be always warm,—and the latent heat given out during the formation of rain and dew. We will briefly consider the manner in which each of these causes contributes to the degree and the uniformity of the equatorial temperature.

Influence of the Heat of the Soil.—It is well known that at a very moderate depth the soil maintains a uniform temperature during the twenty-four hours; while at a greater depth even the annual inequalities disappear, and a uniform temperature, which is almost exactly the mean temperature of the locality, is constantly maintained throughout the year. The depth at which this uniform temperature is reached is greater as the annual range of temperature is greater, so that it is least near the equator, and greatest in localities near the arctic circle where the greatest difference between summer and winter temperature prevails. In the vicinity of the equator, where the annual range of the thermometer is so small as we have seen that it is at Batavia, the mean temperature of about 80° Fahr. is reached at a depth of four or five feet. The surplus heat received during the day is therefore conducted downwards very slowly, the surface soil becomes greatly superheated, and a large portion of this heat is given out at night and thus keeps up the high temperature of the air when the sun has ceased to warm the earth. In the temperate zones, on the other hand, the stratum of uniform earth-temperature lies very deep. At Geneva it is not less than from thirty to forty feet, and with us it is probably fifty or sixty feet, and the temperature found there is nearly forty degrees lower than at the equator. This great body of cool earth absorbs a large portion of the surface heat during the summer, and conducts it downwards with comparative rapidity, and it is only late in the year (in July and August) when the upper layers of the soil have accumulated a surplus store of solar heat that a sufficient quantity is radiated at night to keep up a high temperature in the absence of the sun. At the equator, on the other hand, this radiation is always going on, and earth-heat is one of the most important of the agencies which tend to equalise the equatorial climate.

Influence of the Aqueous Vapour of the Atmosphere. —The aqueous vapour which is always present in considerable quantities in the atmosphere, exhibits a singular and very important relation to solar and terrestrial heat. The rays of the sun pass through it unobstructed to the earth ; but the warmth given off by the heated earth is very largely absorbed by it, thus raising the temperature of the air ; and as it is the lower strata of air which contain most vapour these act as a blanket to the earth, preventing it from losing heat at night by radiation into space. During a large part of the year the air in the equatorial zone is nearly saturated with vapour, so that, notwithstanding the heat, salt and sugar become liquid, and all articles of iron get thickly coated with rust. Complete saturation being represented by 100, the daily average of greatest humidity at Batavia reaches 96 in January and 92 in December. In January, which is the dampest month, the range of humidity is small (77 to 96), and at this time the range of temperature is also least ; while in September, with a greater daily range of humidity (62 to 92) the range of temperature is the greatest, and the lowest temperatures are recorded in this and the preceding month. It is a curious fact, that in many parts of England the degree of humidity as measured by the comparative saturation of the air, is as great as that of Batavia or even greater. A register kept at Clifton during the years 1853—1862 shows a mean humidity in January of 90, while the highest monthly mean for the four years at Batavia was 88 ; and while the lowest of the monthly means at Clifton was 79·1, the lowest at Batavia was 78·9. These figures however represent an immense difference in the *quantity* of vapour in every cubic foot of air. In January at Clifton, with a temperature of 35° to 40° Fahr., there would be only about 4 to 4½ grains of vapour per cubic foot of air, while at Batavia, with a temperature from 80° to 90° Fahr., there would be about 20 grains in the same quantity of air. The most important fact however is, that the capacity of air for holding vapour in suspension increases more rapidly than temperature increases, so that a fall of ten degrees at 50° Fahr. will lead to the condensation of about 1½ grains of vapour, while a similar fall at 90° Fahr. will set free 6½ grains. We can thus understand how it is that the very moderate fall of the thermometer during a tropical night causes heavier dews and a greater amount of sensible moisture than are ever experienced during much greater variations of temperature in the temperate zone. It is this large quantity of vapour in the equatorial atmosphere that keeps up a genial warmth throughout the night by preventing the radiation into space of the heat absorbed by the surface soil during the day. That this is really the case is strikingly proved by what occurs in the plains of Northern India, where the daily maximum of heat is far beyond anything experienced near the equator, yet, owing to the extreme dryness of the atmosphere, the clear nights are very cold, radiation being sometimes so rapid that water placed in shallow pans becomes frozen over.

As the heated earth, and everything upon its surface, does not cool so fast when surrounded by moist as by dry air, it follows, that even if the quantity and intensity of the solar rays falling upon two given portions of the earth's surface are exactly equal, yet the sensible and effective heat produced in the two localities may be very different according as the atmosphere contains much or little vapour. In the one case the heat is absorbed more rapidly than it can escape by radiation ; in the other case it radiates away into space, and is lost, more rapidly than it is being absorbed. In both cases an equilibrium will be arrived at, but in the one case the resulting mean temperature will be much higher than in the other.

Influence of Winds on the Temperature of the Equator. — The distance from the northern to the southern tropics being considerably more than three thousand miles, and the area of the intertropical zone more than one-third the whole area of the globe, it becomes hardly possible for any currents of air to reach the equatorial belt without being previously warmed by contact with the earth or ocean, or by mixture with the heated surface-air which is found in all intertropical and sub-tropical lands. This warming of the air is rendered more certain and more effective by the circumstance, that all currents of air coming from the north or south have their direction changed owing to the increasing rapidity of the earth's rotational velocity, so that they reach the equator as easterly winds, and thus pass obliquely over a great extent of the heated surface of the globe. The causes that produce the westerly monsoons act in a similar manner, so that on the equator direct north or south winds, except as local land and sea breezes, are almost unknown. The Batavia observations show, that for ten months in the year the average direction of the wind varies only between 5° and 30° from due east or west, and these are also the strongest winds. In the two months—March and October—when the winds are northerly, they are very light, and are probably in great part local sea-breezes, which, from the position of Batavia, must come from the north. As a rule, therefore, every current of air at or near the equator has passed obliquely over an immense extent of tropical surface and is thus necessarily a warm wind.

In the north temperate zone, on the other hand, the winds are always cool, and often of very low temperature even in the height of summer, due probably to their coming from colder northern regions as easterly winds, or from the upper parts of the atmosphere as westerly winds; and this constant supply of cool air, combined with quick radiation through a dryer atmosphere, carries off the solar heat so rapidly that an equilibrium is only reached at a comparatively low temperature. In the equatorial zone, on the contrary, the heat accumulates, on account of the absence of any medium of sufficiently low temperature to carry it off rapidly, and it thus soon reaches a point high enough to produce those scorching effects which are so puzzling when the altitude of the sun or the indications of the thermometer are alone considered. Whenever, as is sometimes the case, exceptional cold occurs near the equator, it can almost always be traced to the influence of currents of air of unusually low temperature. Thus in July near the Aru islands, the writer experienced a strong south-east wind which almost neutralised the usual effects of tropical heat although the weather was bright and sunny. But the wind, coming direct from the southern ocean during its winter without acquiring heat by passing over land, was of an unusually low temperature. Again, Mr. Bates informs us that in the Upper Amazon in the month of May there is a regularly recurring south wind which produces a remarkable lowering of the usual equatorial temperature. But owing to the increased velocity of the earth's surface at the equator a south wind there must have been a south-west wind at its origin, and this would bring it directly from the high chain of the Peruvian Andes during the winter of the southern hemisphere. It is therefore probably a cold mountain wind, and blowing as it does over a continuous forest it has been unable to acquire the usual tropical warmth.

The cause of the striking contrast between the climates of equatorial and temperate lands at times when both are receiving an approximately equal amount of solar heat may perhaps be made clearer by an illustration. Let us suppose there to be two reservoirs of water, each supplied by a pipe which pours into it a thousand gallons a day, but which runs only during the daytime, being cut off at night. The reservoirs are both leaky, but while the one loses at the rate of nine hundred gallons in the twenty-four hours the other loses at the rate of eleven hundred gallons in the same time, supposing that both are kept exactly half full and thus subjected to the same uniform water-pressure. If now both are left to be supplied by the above-mentioned pipes the result will be, that in the one which loses by leakage less than it receives the water will rise day by day, till the increased pressure caused the leakage to increase so as exactly to balance the supply ; while in the other the water will sink till the decreasing pressure causes the leakage to decrease so as to balance the supply, when both will remain stationary, the one at a high the other at a low average level, each rising during the day and sinking again at night. Just the same thing occurs with that great heat-reservoir the earth, whose actual temperature at any spot will depend, not alone upon the quantity of heat it receives, but on the balance between its constantly varying waste and supply. We can thus understand how it is that, although in the months of June and July Scotland in latitude 57° north receives as much sun-heat as Angola or Timor in latitude 10° south, and for a much greater number of hours daily, yet in the latter the mean temperature will be about 80° Fahr., with a daily maximum of 90° to 95°, while in the former the mean will be about 60° Fahr. with a daily maximum of 70° or 75° ; and, while in Scotland exposure to the full noon-day sun produces no unpleasant heat-sensations, a similar exposure in Timor at any time between 9 A.M. and 3 P.M. would blister the skin in a few minutes almost as effectually as the application of scalding water.

Heat Due to the Condensation of Atmospheric Vapour.—Another cause which tends to keep up a uniform high temperature in the equatorial, as compared with the variable temperatures of the extra-tropical zones, is the large amount of heat liberated during the condensation of the aqueous vapour of the atmosphere in the form of rain and dew. Owing to the frequent near approach of the equatorial atmosphere to the saturation point, and the great weight of vapour its high temperature enables it to hold in suspension, a very slight fall of the thermometer is accompanied by the

condensation of a large absolute quantity of atmospheric vapour, so that copious dews and heavy showers of rain are produced at comparatively high temperatures and low altitudes. The drops of rain rapidly increase in size while falling through the saturated atmosphere ; and during this process as well as by the formation of dew, the heat which retained the water in the gaseous form, and was insensible while doing so, is liberated, and thus helps to keep up the high temperature of the air. This production of heat is almost always going on. In fine weather the nights are always dewy, and the diagram on the preceding page showing the mean monthly rainfall at Batavia and Greenwich proves that this source of increased temperature is present during every month in the year, since the lowest monthly fall at the former place is almost equal to the highest monthly fall at the latter.

It may perhaps be objected, that evaporation must absorb as much heat as is afterwards liberated by condensation, and this is true ; but as evaporation and condensation occur usually at different times and in different places, the equalising effect is still very important. Evaporation occurs chiefly during the hottest sunshine, when it tends to moderate the extreme heat, while condensation takes place chiefly at night in the form of dew and rain, when the liberated heat helps to make up for the loss of the direct rays of the sun. Again, the most copious condensation both of dew and rain is greatly influenced by vegetation and especially by forests, and also by the presence of hills and mountains, and is therefore greater on land than on the ocean ; while evaporation is much greater on the ocean, both on account of the less amount of cloudy weather and because the air is more constantly in motion. This is particularly the case throughout that large portion of the tropical and subtropical zones where the trade-winds constantly blow, as the evaporation must there be enormous while the quantity of rain is very small. It follows, then, that on the equatorial land-surface there will be a considerable balance of condensation over evaporation which must tend to the general raising of the temperature, and, owing to the condensation being principally at night, not less powerfully to its equalisation.

General Features of the Equatorial Climate.—The various causes now enumerated are sufficient to enable us to understand how the great characteristic features of the climate of the equatorial zone are brought about ; how it is that so high a temperature is maintained during the absence of the sun at night, and why so little effect is produced by the sun's varying altitude during its passage from the northern to the southern tropic. In this favoured zone the heat is never oppressive, as it so often becomes on the borders of the tropics ; and the large absolute amount of moisture always present in the air, is almost as congenial to the health of man as it is favourable to the growth and development of vegetation.[1] Again, the lowering of the temperature at night is so regular and yet so strictly limited in amount, that, although never cold enough to be unpleasant, the nights are never so oppressively hot as to prevent sleep. During the wettest months of the year, it is rare to have many days in succession

---

[1] Where the inhabitants adapt their mode of life to the peculiarities of the climate, as is the case with the Dutch in the Malay Archipelago, they enjoy as robust health as in Europe, both in the case of persons born in Europe and of those who for generations have lived under a vertical sun.

C

without some hours of sunshine, while even in the driest months there are occasional showers to cool and refresh the overheated earth. As a result of this condition of the earth and atmosphere, there is no check to vegetation, and little if any demarcation of the seasons. Plants are all evergreen ; flowers and fruits, although more abundant at certain seasons, are never altogether absent ; while many annual food-plants as well as some fruit-trees produce two crops a year. In other cases, more than one complete year is required to mature the large and massive fruits, so that it is not uncommon for fruit to be ripe at the same time that the tree is covered with flowers, in preparation for the succeeding crop. This is the case with the Brazil nut tree, in the forests of the Amazon, and with many other tropical as with a few temperate fruits.

Uniformity of the Equatorial Climate in all Parts of the Globe.—The description of the climatal phenomena of the equatorial zone here given, has been in great part drawn from long personal experience in South America and in the Malay Archipelago. Over a large portion of these countries the same general features prevail, only modified by varying local conditions. Whether we are at Singapore or Batavia ; in the Moluccas, or New Guinea ; at Para, at the sources of the Rio Negro, or on the Upper Amazon, the equatorial climate is essentially the same, and we have no reason to believe that it materially differs in Guinea or the Congo. In certain localities, however, a more contrasted wet and dry season prevails, with a somewhat greater range of the thermometer. This is generally associated with a sandy soil, and a less dense forest, or with an open and more cultivated country. The open sandy country with scattered trees and shrubs or occasional thickets, which is found at Santarem and Monte-Alegre on the lower Amazon, are examples, as well as the open cultivated plains of Southern Celebes ; but in both cases the forest country in adjacent districts has a moister and more uniform climate, so that it seems probable that the nature of the soil or the artificial clearing away of the forests, are important agents in producing the departure from the typical equatorial climate observed in such districts. The almost rainless district of Ceara on the North-East coast of Brazil and only a few degrees south of the equator, is a striking example of the need of vegetation to react on the rainfall. We have here no apparent cause but the sandy soil and bare hills, which when heated by the equatorial sun produce ascending currents of warm air and thus prevent the condensation of the atmospheric vapour, to account for such an anomaly ; and there is probably no district where judicious planting would produce such striking and beneficial effects. In Central India the scanty and intermittent rainfall, with its fearful accompaniment of famine, is no doubt in great part due to the absence of a sufficient proportion of forest-covering to the earth's surface ; and it is to a systematic planting of all the hill tops, elevated ridges, and higher slopes that we can alone look for a radical cure of the evil. This would almost certainly induce an increased rainfall ; but even more important and more certain, is the action of forests in checking evaporation from the soil and causing perennial springs to flow, which may be collected in vast storage tanks and will serve to fertilise a great extent of country ; whereas tanks without regular rainfall or permanent springs to supply them are worthless. In the colder parts of the temperate zones, the absence of forests is not so much felt, because the hills and uplands are naturally clothed with a thick coating of turf which absorbs moisture and does not become overheated by the sun's rays, and the rains are seldom violent enough to strip this protective covering from the surface. In tropical and even in south-temperate countries, on the other hand, the rains are periodical and often of excessive violence for a short period ; and when the forests are cleared away the torrents of rain soon strip off the vegetable soil, and thus destroy in a few years the fertility which has been the growth of many centuries. The bare subsoil becoming heated by the sun, every particle of moisture which does not flow off is evaporated, and this again reacts on the climate, producing long-continued droughts only relieved by sudden and violent storms, which add to the destruction and render all attempts at cultivation unavailing. Wide tracts of fertile land in the south of Europe have been devastated in this manner, and have become absolutely uninhabitable. Knowingly to produce such disastrous results would be a far more serious offence than any destruction of property which human labour has produced and can replace ; yet we ignorantly allow such extensive clearings for coffee cultivation in India and Ceylon, as to cause the destruction of much fertile soil which generations cannot replace, and which will surely, if not checked in time, lead to the deterioration of the climate and the permanent impoverishment of the country.[1]

Short Twilight of the Equatorial Zone.—One of the phenomena which markedly distinguish the equatorial from the temperate and polar zones, is the shortness of the twilight and consequent rapid transition from day to night and from night to day. As this depends only on the fact of the sun descending vertically instead of obliquely below the horizon, the difference is most marked when we compare our midsummer twilight with that of the tropics. Even with us the duration of twilight is very much shorter at the time of the equinoxes, and it is probably not much more than a third shorter than this at the equator. Travellers usually exaggerate the shortness of the tropical twilight, it being sometimes said that if we turn a page of the book we are reading when the sun disappears, by the time we turn over the next page it will be too dark to see to read. With an average book and an average reader this is certainly not true, and it will be well to describe as correctly as we can what really happens.

In fine weather the air appears to be somewhat more transparent near the equator than with us, and the intensity of sunlight is usually very great up to the moment when the solar orb touches the horizon. As soon as it has disappeared the apparent gloom is proportionally great, but this hardly increases perceptibly during the first ten minutes. During the next ten minutes however it becomes rapidly darker, and at the end of

---

[1] For a terrible picture of the irreparable devastation caused by the reckless clearing of forests see the third chapter of Mr. Marsh's work *The Earth as Modified by Human Action.*

about twenty-five minutes from sunset the complete darkness of night is almost reached. In the morning the changes are perhaps even more striking. Up to about a quarter past five o'clock the darkness is complete; but about that time a few cries of birds begin to break the silence of night, perhaps indicating that signs of dawn are perceptible in the eastern horizon. A little later the melancholy voices of the goatsuckers are heard, varied croakings of frogs, the plaintive whistle of mountain thrushes, and strange cries of birds or mammals peculiar to each locality. About half-past five the first glimmer of light becomes perceptible; it slowly becomes lighter, and then increases so rapidly that at about a quarter to six it seems full daylight. For the next quarter of an hour this changes very little in character; when, suddenly, the sun's rim appears above the horizon, decking the dew-laden foliage with glittering gems, sending gleams of golden light far into the woods, and waking up all nature to life and activity. Birds chirp and flutter about, parrots scream, monkeys chatter, bees hum among the flowers, and gorgeous butterflies flutter lazily along or sit with fully expanded wings exposed to the warm and invigorating rays. The first hour of morning in the equatorial regions possesses a charm and a beauty that can never be forgotten. All nature seems refreshed and strengthened by the coolness and moisture of the past night; new leaves and buds unfold almost before the eye, and fresh shoots may often be observed to have grown many inches since the preceding day. The temperature is the most delicious conceivable. The slight chill of early dawn, which was itself agreeable, is succeeded by an invigorating warmth; and the intense sunshine lights up the glorious vegetation of the tropics, and realises all that the magic art of the painter or the glowing words of the poet, have pictured as their ideals of terrestrial beauty.

The Aspect of the Equatorial Heavens.—Within the limits of the equatorial zone the noonday sun is truly vertical twice every year, and for several months it passes so near the zenith that the difference can hardly be detected without careful observation of the very short shadows of vertical objects. The absence of distinct horizontal shadows at noon which thus characterises a considerable part of the year, is itself a striking pheno- menon to an inhabitant of the temperate zones ; and equally striking is the changed aspect of the starry heavens. The grand constellation Orion, passes verti- cally overhead, while the Great Bear is only to be seen low down in the northern heavens, and the Pole star either appears close to the horizon or has altogether disappeared according as we are north or south of the equator. Towards the south the Southern Cross, the Magellanic clouds, and the jet-black " coal sacks " are the most conspicuous objects invisible in our northern lati- tudes. The same cause that brings the sun overhead in its daily march equally affects the planets, which appear high up towards the zenith far more frequently than with us, thus affording splendid opportunities for telescopic observation.

Intensity of Meteorological Phenomena at the Equator. —The excessive violence of meteorological phenomena generally supposed to be characteristic of the tropics is not by any means remarkable in the equatorial zone. Electrical disturbances are much more frequent, but not generally more violent than in the temperate regions. The wind-storms are rarely of excessive violence, as might in fact be inferred from the extreme steadiness of the barometer, whose daily range at Batavia rarely exceeds one-eighth of an inch, while the extreme range during three years was less than one-third of an inch! The amount of the rainfall is very great, seventy or eighty inches in a year being a probable average ; and as the larger part of this occurs during three or four months, individual rainfalls are often exceedingly heavy. The greatest fall recorded at Batavia during three years was three inches and eight-tenths in one hour,[1] but this was quite exceptional, and even half this quantity is very unusual. The greatest rainfall recorded in twenty-four hours is seven inches and a quarter ; but more than four inches in one day occurs only on two or three occasions in a year. The blue colour of the sky is probably not so intense as in many parts of the temperate zone, while the brilliancy of the moon and stars is not perceptibly greater than that of our clearest frosty nights, and is undoubtedly much inferior to what is witnessed in many desert regions, and even in Southern Europe.

On the whole, then, we must decide, that uniformity and abundance, rather than any excessive manifestations, are the prevailing characteristic of all the climatal phenomena of the equatorial zone.

Concluding Remarks.—We cannot better conclude our account of the equatorial climate than by quoting the following vivid description of the physical phenomena which occur during the early part of the dry season at Para. It is taken from Mr. Bates' *Naturalist*

[1] On January 10th, 1867, from 1 to 2 A.M.

on the Amazons, and clearly exhibits some of the more characteristic features of a typical equatorial day.

"At that early period of the day (the first two hours after sunrise) the sky was invariably cloudless, the thermometer marking 72° or 73° Fahr.; the heavy dew or the previous night's rain, which lay on the moist foliage, becoming quickly dissipated by the glowing sun, which, rising straight out of the east, mounted rapidly towards the zenith. All nature was fresh, new leaf and flower-buds expanding rapidly. * * * The heat increased hourly, and towards two o'clock reached 92° to 93° Fahr., by which time every voice of bird and mammal was hushed. The leaves, which were so moist and fresh in early morning, now became lax and drooping, and flowers shed their petals. On most days in June and July a heavy shower would fall some time in the afternoon, producing a most welcome coolness. The approach of the rain-clouds was after a uniform fashion very interesting to observe. First, the cool sea-breeze which had commenced to blow about ten o'clock, and which had increased in force with the increasing power of the sun, would flag, and finally die away. The heat and electric tension of the atmosphere would then become almost insupportable. Languor and uneasiness would seize on every one, even the denizens of the forest betraying it by their motions. White clouds would appear in the east and gather into cumuli, with an increasing blackness along their lower portions. The whole eastern horizon would become almost suddenly black, and this would spread upwards, the sun at length becoming obscured. Then the rush of a mighty wind is heard through the forest, swaying the tree-tops ; a vivid flash of lightning bursts forth, then a crash of thunder, and down streams the deluging rain. Such storms soon cease, leaving bluish-black motionless clouds in the sky until night. Meantime all nature is refreshed ; but heaps of flower-petals and fallen leaves are seen under the trees. Towards evening life revives again, and the ringing uproar is resumed from bush and tree. The following morning the sun again rises in a cloudless sky ; and so the cycle is completed ; spring, summer, and autumn, as it were in one tropical day. The days are more or less like this throughout the year. A little difference exists between the dry and wet seasons; but generally, the dry season, which lasts from July to December, is varied with showers, and the wet, from January to June, with sunny days. It results from this, —that the periodical phenomena of plants and animals do not take place at about the same time in all species, or in the individuals of any given species, as they do in temperate countries. In Europe, a woodland scene has its spring, its summer, its autumnal, and its winter aspects. In the equatorial forests the aspect is the same or nearly so every day in the year : budding, flowering, fruiting, and leaf-shedding are always going on in one species or other It is never either spring, summer, or autumn, but each day is a combination of all three. With the day and night always of equal length, the atmospheric disturbances of each day neutralising them-selves before each succeeding morn ; with the sun in its course proceeding midway across the sky, and the daily temperature almost the same throughout the year—how grand in its perfect equilibrium and simplicity is the march of Nature under the equator ! ''

II.

EQUATORIAL VEGETATION.

The Equatorial Forest-Belt and its Causes—General features of the Equatorial Forests—Low-growth Forest-trees—Flowery trunks and their probable cause—Uses of Equatorial Forest-trees—The Climbing Plants of the Equatorial Forests—Palms—Uses of Palm-trees and their Products— Ferns—Ginger-worts and wild Bananas—Arums—Screw-pines—Orchids —Bamboos — Uses of the Bamboo — Mangroves—Sensitive-plants— Comparative scarcity of Flowers—Concluding Remarks on Tropical Vegetation.

In the following sketch of the characteristics of vegetable life in the equatorial zone, it is not intended to enter into any scientific details or to treat the subject in the slightest degree from a botanical point of view; but merely to describe those general features of vegetation which are almost or quite peculiar to this region of the globe, and which are so general as to be characteristic of the greater part of it rather than of any particular country or continent within its limits.

The Equatorial Forest-Belt and its Causes.—With but few and unimportant exceptions a great forest band from a thousand to fifteen hundred miles in width girdles the earth at the equator, clothing hill, plain, and mountain with an evergreen mantle. Lofty peaks and precipitous ridges are sometimes bare, but often the woody covering continues to a height of eight or ten

thousand feet, as in some of the volcanic mountains
of Java and on portions of the Eastern Andes. Beyond
the forests both to the north and south, we meet first
with woody and then open country, soon changing into
arid plains or even deserts which form an almost continuous band in the vicinity of the two tropics. On
the line of the tropic of Cancer we have, in America
the deserts and dry plains of New Mexico; in Africa the
Sahara; and in Asia, the Arabian deserts, those of Beloo-
chistan and Western India, and further east the dry
plains of North China and Mongolia. On the tropic of
Capricorn we have, in America the Grand Chaco desert
and the Pampas; in Africa the Kalahari desert and the
dry plains north of the Limpopo; while the deserts and
waterless plains of Central Australia complete the arid zone.
These great contrasts of verdure and barrenness occurring
in parallel bands all round the globe, must evidently
depend on the general laws which determine the distribution of moisture over the earth, more or less modified
by local causes. Without going into meteorological
details, some of which have been given in the preceding
chapter, the main facts may be explained by the mode
in which the great aerial currents are distributed. The
trade winds passing over the ocean from north-east to
south-west with an oblique tendency towards the equator,
become saturated with vapour, and are ready to give
out moisture whenever they are forced upwards or in any
other way have their temperature lowered. The entire
equatorial zone becomes thus charged with vapour-laden
air which is the primary necessity of a luxuriant vegetation. The surplus air (produced by the meeting of the
two trade winds) which is ever rising in the equatorial

belt and giving up its store of vapour, flows off north and south as dry, cool air, and descends to the earth in the vicinity of the tropics. Here it sucks up whatever moisture it meets with and thus tends to keep this zone in an arid condition. The trades themselves are believed to be supplied by descending currents from the temperate zones, and these are at first equally dry and only become vapour-laden when they have passed over some extent of moist surface. At the solstices the sun passes vertically over the vicinity of the tropics for several weeks, and this further aggravates the aridity ; and wherever the soil is sandy and there are no lofty mountain-chains to supply ample irrigation the result is a more or less perfect desert. / Analogous causes, which a study of aerial currents will render intelligible, have produced other great forest-belts in the northern and southern parts of the temperate zones ; but owing to the paucity of land in the southern hemisphere these are best seen in North America and Northern Euro-Asia, where they form the great northern forests of deciduous trees and of Coniferæ. These being comparatively well-known to us, will form the standard by a reference to which we shall endeavour to point out and render intelligible the distinctive characteristics of the equatorial forest vegetation.

General Features of the Equatorial Forests.—It is not easy to fix upon the most distinctive features of these virgin forests, which nevertheless impress themselves upon the beholder as something quite unlike those of temperate lands, and as possessing a grandeur and sublimity altogether their own. Amid the countless modifications in detail which these forests present, we shall endeavour to point out the chief peculiarities as well as the more interesting phenomena which generally characterise them.

The observer new to the scene would perhaps be first struck by the varied yet symmetrical trunks, which rise up with perfect straightness to a great height without a branch, and which, being placed at a considerable average distance apart, give an impression similar to that produced by the columns of some enormous building. Overhead, at a height, perhaps, of a hundred feet, is an almost unbroken canopy of foliage formed by the meeting together of these great trees and their interlacing branches; and this canopy is usually so dense that but an indistinct glimmer of the sky is to be seen, and even the intense tropical sunlight only penetrates to the ground subdued and broken up into scattered fragments. There is a weird gloom and a solemn silence, which combine to produce a sense of the vast—the primeval—almost of the infinite. It is a world in which man seems an intruder, and where he feels overwhelmed by the contemplation of the ever-acting forces, which, from the simple elements of the atmosphere, build up the great mass of vegetation which overshadows, and almost seems to oppress the earth.

Characteristics of the Larger Forest-trees.—Passing from the general impression to the elements of which the scene is composed, the observer is struck by the great diversity of the details amid the general uniformity. Instead of endless repetitions of the same forms of trunk such as are to be seen in our pine, or oak, or beech woods, the eye wanders from one tree to another and rarely detects two of the same species. All are tall and upright columns, but they differ from each other more than do the columns of Gothic, Greek, and Egyptian temples. Some are almost cylindrical, rising up out of the ground as if their bases were concealed by accumulations of the soil ; others get much thicker near the ground like our spreading oaks; others again, and these are very characteristic, send out towards the base flat and wing-like projections. These projections are thin slabs radiating from the main trunk, from which they stand out like the buttresses of a Gothic cathedral. They rise to various heights on the tree, from five or six, to twenty or thirty feet ; they often divide as they approach the ground, and sometimes twist and curve along the surface for a considerable distance, forming elevated and greatly compressed roots. These buttresses are sometimes so large that the spaces between them if roofed over would form huts capable of containing several persons. Their use is evidently to give the tree an extended base, and so assist the subterranean roots in maintaining in an erect position so lofty a column crowned by a broad and massive head of branches and foliage. The buttressed trees belong to a variety of distinct groups. Thus, many of the Bombaceæ or silk-cotton trees, several of the Leguminosæ, and perhaps many trees belonging to other natural orders, possess these appendages.

There is another form of tree, hardly less curious, in which the trunk, though generally straight and cylindrical, is deeply furrowed and indented, appearing as if made up of a number of small trees grown together at the centre. Sometimes the junction of what seem to be the component parts, is so imperfect, that gaps or holes are left by which you can see through the trunk in various places. At first one is disposed to think this is

caused by accident or decay, but repeated examination
shows it be due to the natural growth of the tree.   The
accompanying outline sections of one of these trees that
was cut down, exhibits its character.   It was a noble

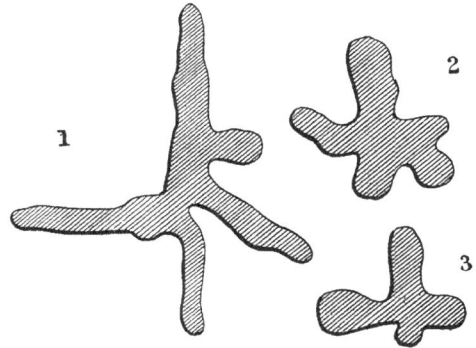

Sections of trunk of a Bornean Forest-tree.
1. Section at seven feet from the ground.
2. 3. Sections much higher up.

forest-tree, more than 200 feet high, but rather slender
in proportion, and it was by no means an extreme
example of its class.   This peculiar form is probably
produced by the downward growth of aerial roots, like
some New Zealand trees whose growth has been traced,
and of whose different stages drawings may be seen at
the Library of the Linnean Society.   These commence
their existence as parasitical climbers which take root in
the fork of some forest-tree and send down aerial roots
which clasp round the stem that upholds them.   As
these roots increase in size and grow together laterally
they cause the death of their foster-parent.   The climber
then grows rapidly, sending out large branches above
and spreading roots below, and as the supporting tree
decays away the aerial roots grow together and form a

new trunk, more or less furrowed and buttressed, but exhibiting no other marks of its exceptional origin. Aerial-rooted forest-trees—like that figured in my *Malay Archipelago* (vol. i. p. 131)—and the equally remarkable fig-trees of various species, whose trunks are formed by a miniature forest of aerial roots, sometimes separate, sometimes matted together, are characteristic of the Eastern tropics, but appear to be rare or altogether unknown in America, and can therefore hardly be included among the general characteristics of the equatorial zone.

Besides the varieties of form, however, the tree-trunks of these forests present many peculiarities of colour and texture. The majority are rather smooth-barked, and many are of peculiar whitish, green, yellowish, or brown colours, or occasionally nearly black. Some are perfectly smooth, others deeply cracked and furrowed, while in a considerable number the bark splits off in flakes or hangs down in long fibrous ribands. Spined or prickly trunks (except of palms) are rare in the damp equatorial forests. Turning our gaze upwards from the stems to the foliage, we find two types of leaf not common in the temperate zone, although the great mass of the trees offer nothing very remarkable in this respect. First, we have many trees with large, thick, and glossy leaves, like those of the cherry-laurel or the magnolia, but even larger, smoother, and more symmetrical. The leaves of the Asiatic caoutchouc-tree (*Ficus elastica*), so often cultivated in houses, is a type of this class, which has a very fine effect among the more ordinary-looking foliage. Contrasted with this is the fine pinnate foliage of some of the largest forest-trees which, seen far aloft against the sky, looks as delicate as that of the sensitive mimosa.

Forest-trees of Low Growth.—The great trees we have hitherto been describing form, however, but a portion of the forest. Beneath their lofty canopy there often exists a second forest of moderate-sized trees, whose crowns, perhaps forty or fifty feet high, do not touch the lowermost branches of those above them. These are of course shade-loving trees, and their presence effectually prevents the growth of any young trees of the larger kinds, until, overcome by age and storms, some monarch of the forest falls down, and, carrying destruction in its fall, opens up a considerable space, into which sun and air can penetrate. Then comes a race for existence among the seedlings of the surrounding trees, in which a few ultimately prevail and fill up the space vacated by their predecessor. Yet beneath this second set of medium-sized forest-trees there is often a third undergrowth of small trees, from six to ten feet high, of dwarf palms, of tree-ferns, and of gigantic herbaceous ferns. Coming to the surface of the ground itself we find much variety. Sometimes it is completely bare, a mass of decaying leaves and twigs and fallen fruits. More frequently it is covered with a dense carpet of selaginella or other lycopodiaceæ, and these sometimes give place to a variety of herbaceous plants, sometimes with pretty, but rarely with very conspicuous flowers.

Flowering Trunks and their Probable Cause.— Among the minor but not unimportant peculiarities that characterise these lofty forests, is the curious way in which many of the smaller trees have their flowers situated on the main trunk or larger branches instead of on the upper part of the tree. The cacao-tree is a well-known example of this peculiarity, which is not uncommon in tropical forests; and some of the smaller trunks are occasionally almost hidden by the quantity of fruit produced on them. One of the most beautiful examples of this mode of flowering is a small tree of the genus *Polyalthea*, belonging to the family of the custard-apples, not uncommon in the forests of North-western Borneo. Its slender trunk, about fifteen or twenty feet high, was completely covered with star-shaped flowers, three inches across and of a rich orange-red colour, making the trees look as if they had been artificially decorated with brilliant garlands. The recent discoveries as to the important part played by insects in the fertilization of flowers offers a very probable explanation of this peculiarity. Bees and butterflies are the greatest flower-haunters. The former love the sun and frequent open grounds or the flowery tops of the lofty forest-trees fully exposed to the sun and air. The forest shades are frequented by thousands of butterflies, but these mostly keep near the ground, where they have a free passage among the tree-trunks and visit the flowering shrubs and herbaceous plants. To attract these it is necessary that flowers should be low down and conspicuous. If they grew in the usual way on the tops of these smaller trees overshadowed by the dense canopy above them they would be out of sight of both groups of insects, but being placed openly on the stems, and in the greatest profusion, they cannot fail to attract the attention of the wandering butterflies.

Uses of Equatorial Forest-trees.—Amid this immense variety of trees, the natives have found out such as are best adapted to certain purposes. The wood of some is light and soft, and is used for floats or for carving out rude images, stools, and ornaments for boats and houses. The flat slabs of the buttresses are often used to make paddles. Some of the trees with furrowed stems are exceedingly strong and durable, serving as posts for houses or as piles on which the water-villages are built. Canoes, formed from a trunk hollowed out and spread open under the action of heat, require one kind of wood, those built up with planks another; and, as the species of trees in these forests are so much more numerous than the wants of a semi-civilized population, there are probably a large number of kinds of timber which will some day be found to be well adapted to the special requirements of the arts and sciences. The products of the trees of the equatorial forests, notwithstanding our imperfect knowledge of them, are already more useful to civilized man than to the indigenous inhabitants. To mention only a few of those whose names are tolerably familiar to us, we have such valuable woods as mahogany, teak, ebony, lignum-vitæ, purple-heart, iron-wood, sandal-wood, and satin-wood ; such useful gums as india-rubber, gutta-percha, tragacanth, copal, lac, and dammar ; such dyes as are yielded by log-wood, brazil-wood, and sappan-wood ; such drugs as the balsams of Capivi and Tolu, camphor, benzoin, catechu or terra-japonica, caju-put oil, gamboge, quinine, Angostura bark, quassia, and the urari and upas poisons ; of spices we have cloves, cinnamon, and nutmegs ; and of fruits, brazil-nuts, tamarinds, guavas, and the valuable cacao ; while residents in our tropical colonies enjoy the bread-fruit, avocado-pear, custard-apple, durian, mango, mangosteen, soursop, papaw, and many others. This list of useful products from the exogenous trees alone of the equatorial forests, excluding those from the palms, shrubs, herbs, and creepers, might have been multiplied many times over by the introduction of articles whose names would be known only to those interested in special arts or sciences; but imperfect as it is, it will serve to afford a notion of the value of this vast treasure-house which is as yet but very partially explored.

The Climbing Plants of the Equatorial Forests.—Next to the trees themselves the most conspicuous and remarkable feature of the tropical forests is the profusion of woody creepers and climbers that everywhere meet the eye. They twist around the slenderer stems, they drop down pendent from the branches, they stretch tightly from tree to tree, they hang looped in huge festoons from bough to bough, they twist in great serpentine coils or lie in entangled masses on the ground. Some are slender, smooth, and root-like; others are rugged or knotted; often they are twined together into veritable cables ; some are flat like ribands, others are curiously waved and indented. Where they spring from or how they grow is at first a complete puzzle. They pass overhead from tree to tree, they stretch in tight cordage like the rigging of a ship from the top of one tree to the base of another, and the upper regions of the forest often seem full of them without our being able to detect any earth-growing stem from which they arise. The conclusion is at length forced upon us that these woody climbers must possess the two qualities of very long life and almost indefinite longitudinal growth, for by these suppositions alone can we explain their characteristic features. The growth of climbers, even more than all other plants, is upward towards the light. In the shade of the forest they rarely or never flower, and seldom even produce foliage ; but when they have reached the summit of the tree that supports them, they expand under the genial influence of light and air, and often cover their foster-parent with blossoms not its own. Here, as a rule, the climber's growth would cease ; but the time comes when the supporting tree rots and falls, and the creeper comes with it in torn and tangled masses to the ground. But though its foster-parent is dead it has itself received no permanent injury, but shoots out again till it finds a fresh support, mounts another tree, and again puts forth its leaves and flowers. In time the old tree rots entirely away and the creeper remains tangled on the ground. Sometimes branches only fall and carry a portion of the creeper tightly stretched to an adjoining tree ; at other times the whole tree is arrested by a neighbour to which the creeper soon transfers itself in order to reach the upper light. When by the fall of a branch the creepers are left hanging in the air, they may be blown about by the wind and catch hold of trees growing up beneath them, and thus become festooned from one tree to another. When these accidents and changes have been again and again repeated the climber may have travelled very far from its parent stem, and may have mounted to the tree tops and descended again to the earth several times over. Only in this way does it seem possible to explain the wonderfully complex manner in which these climbing plants wander up and down the forest as if guided by the strangest caprices, or how they become so crossed and tangled together in the wildest confusion.

The variety in the length, thickness, strength and toughness of these climbers, enables the natives of tropical countries to put them to various uses. Almost every kind of cordage is supplied by them. Some will stand in water without rotting, and are used for cables, for lines to which are attached fish-traps, and to bind and strengthen the wooden anchors used generally in the East. Boats and even large sailing vessels are built, whose planks are entirely fastened together by this kind of cordage skilfully applied to internal ribs. For the better kinds of houses, smooth and uniform varieties are chosen, so that the beams and rafters can be bound together with neatness, strength and uniformity, as is especially observable among the indigenes of the Amazonian forests. When baskets of great strength are required special kinds of creepers are used ; and to serve almost every purpose for which we should need a rope or a chain, the tropical savage adopts some one of the numerous forest-ropes which long experience has shown to have qualities best adapted for it. Some are smooth and supple ; some are tough and will bear twisting or tying ; some will last longest in salt water, others in fresh ; one is uninjured by the heat and smoke of fires, while another is bitter or otherwise prejudicial to insect enemies.

Besides these various kinds of trees and climbers which form the great mass of the equatorial forests and determine their general aspect, there are a number of forms of plants which are always more or less present, though in some parts scarce and in others in great profusion, and which largely aid in giving a special character to tropical as distinguished from temperate vegetation. Such are the various groups of palms, ferns, ginger-worts, and wild plantains, arums, orchids, and bamboos; and under these heads we shall give a short account of the part they take in giving a distinctive aspect to the equatorial forests.

Palms. — Although these are found throughout the tropics and a few species even extend into the warmer parts of the temperate regions, they are yet so much more abundant and varied within the limits of the region we are discussing that they may be considered as among the most characteristic forms of vegetation of the equatorial zone. They are, however, by no means generally present, and we may pass through miles of forest without even seeing a palm. In other parts they abound; either forming a lower growth in the lofty forest, or in swamps and on hill-sides sometimes rising up above the other trees. On river-banks they are especially conspicuous and elegant, bending gracefully over the stream, their fine foliage waving in the breeze, and their stems often draped with hanging creepers.

The chief feature of the palm tribe consists in the cylindrical trunk crowned by a mass of large and somewhat rigid leaves. They vary in height from a few feet to that of the loftiest forest-trees. Some are stemless, consisting only of a spreading crown of large pinnate leaves; but the great majority have a trunk slender in proportion to its height. Some of the smaller species have stems no thicker than a lead pencil, and four or five feet high; while the great Mauritia of the Amazon has a trunk full two feet in diameter, and more than 100 feet high. Some species probably reach a height of 200 feet, for Humboldt states that in South America he measured a palm, which was 192 English feet high. The leaves of palms are often of immense size. Those of the *Manicaria saccifera* of Para are thirty feet long and four or five feet wide, and are not pinnate but entire and very rigid. Some of the pinnate leaves are much larger, those of the *Raphia tædigera* and *Maximiliana regia* being both sometimes more than fifty feet long. The fan-shaped leaves of other species are ten or twelve feet in diameter. The trunks of palms are sometimes smooth and more or less regularly ringed, but they are frequently armed with dense prickles which are sometimes eight inches long. In some species, the leaves fall to the ground as they decay leaving a clean scar, but in most cases they are persistent, rotting slowly away, and leaving a mass of fibrous stumps attached to the upper part of the stem. This rotting mass forms an excellent soil for ferns, orchids, and other semi-parasitical plants, which form an attractive feature on what would otherwise be an unsightly object. The sheathing margins of the leaves often break up into a fibrous material, sometimes resembling a coarse cloth, and in other cases more like horsehair. The flowers are not individually large, but form large spikes or racemes, and the fruits are often beautifully scaled and hang in huge bunches which are sometimes more than a load for a strong man. The climbing palms are very remarkable, their tough, slender, prickly stems mounting up by means of the hooked midribs of the leaves to the tops of the loftiest forest-trees, above which they send up an elegant spike of foliage and flowers. The most important are the American *Desmoncus* and the Eastern *Calamus*, the

latter being the well-known rattan or cane of which
chair-seats are made, from the Malay name "rotang."
The rattan-palms are the largest and most remarkable
of the climbing group. They are very abundant in the
drier equatorial forests, and more than sixty species are
known from the Malay Archipelago. The stems (when
cleaned from the sheathing leaves and prickles) vary in
size from the thickness of a quill to that of the wrist;
and where abundant they render the forest almost im-
passable. They lie about the ground coiled and twisted
and looped in the most fantastic manner. They hang
in festoons from trees and branches, they rise suddenly
through mid air up to the top of the forest, or coil
loosely over shrubs and in thickets like endless serpents.
They must attain an immense age, and apparently
have almost unlimited powers of growth, for some are
said to have been found which were 600 or even 1000
feet long, and if so, they are probably the longest of all
vegetable growths. The mode in which such great
lengths and tangled convolutions have been attained
has already been explained in the general account of
woody climbers. From the immense strength of these
canes and the facility with which they can be split, they
are universally used for cordage in the countries where
they grow in preference to any other climbers, and
immense quantities are annually exported to all parts of
the world.

Uses of Palm-trees and their Products.—To the
natives of the equatorial zone the uses of palms are
both great and various. The fruits of several species—
more especially the cocoa-nut of the East and the
peach-nut (*Guilielma speciosa*) of America—furnish

abundance of wholesome food, and the whole of the trunk of the sago-palm is converted into an edible starch— our sago. Many other palm-fruits yield a thin pulp, too small in quantity to be directly eaten, but which when rubbed off and mixed with a proper quantity of water forms an exceedingly nutritious and agreeable article of food. The most celebrated of these is the assai of the Amazon, made from the fruit of *Euterpe oleracea*, and which, as a refreshing, nourishing, and slightly stimulating beverage for a tropical country, takes the place of our chocolate and coffee. A number of other palms yield a similar product, and many that are not eaten by man are greedily devoured by a variety of animals, so that the amount of food produced by this tribe of plants is much larger than is generally supposed.

The sap which pours out of the cut flower-stalk of several species of palm when slightly fermented forms palm-wine or toddy, a very agreeable drink ; and when mixed with various bitter herbs or roots which check fermentation, a fair imitation of beer is produced. If the same fluid is at once boiled and evaporated it produces a quantity of excellent sugar. The *Arenga saccharifera*, or sugar-palm of the Malay countries, is perhaps the most productive of sugar. A single tree will continue to pour out several quarts of sap daily for weeks together, and where the trees are abundant this forms the chief drink and most esteemed luxury of the natives. A Dutch chemist, Mr. De Vry, who has studied the subject in Java, believes that great advantages would accrue from the cultivation of this tree in place of the sugar-cane. According to his experiments it would produce an equal quantity of sugar of good quality with far less labour and expense, because no manure and no cultivation would be required, and the land will never be impoverished as it so rapidly becomes by the growth of sugar-cane. The reason of this difference is, that the whole produce of a cane-field is taken off the ground, the crushed canes being burnt; and the soil thus becomes exhausted of the various salts and minerals which form part of the woody fibre and foliage. These must be restored by the application of manure, and this, together with the planting, weeding, and necessary cultivation, is very expensive. With the sugar-palm, however, nothing whatever is taken away but the juice itself; the foliage falls on the ground and rots, giving back to it what it had taken; and the water and sugar in the juice being almost wholly derived from the carbonic acid and aqueous vapour of the atmosphere, there is no impoverishment; and a plantation of these palms may be kept up on the same ground for an indefinite period. Another most important consideration is, that these trees will grow on poor rocky soil and on the steep slopes of ravines and hillsides where any ordinary cultivation is impossible, and a great extent of fertile land would thus be set free for other purposes. Yet further, the labour required for such sugar plantations as these would be of a light and intermittent kind, exactly suited to a semi-civilized people to whom severe and long-continued labour is never congenial. This combination of advantages appears to be so great, that it seems possible that the sugar of the world may in the future be produced from what would otherwise be almost waste ground; and it is to be hoped that the experiment will soon be tried in some of our tropical colonies, more especially as an Indian palm, *Phœnix sylvestris,* also produces abundance of sugar, and might be tried in its native country.

Other articles of food produced from palms are, cooking-oil from the cocoa-nut and baccaba palm, salt from the fruit of a South American palm (*Leopoldinia major*), while the terminal bud or " cabbage" of many species is an excellent and nutritious vegetable ; so that palms supply bread, oil, sugar, salt, fruit, and vegetables. Oils for various other purposes are made from several distinct palms, while wax is secreted from the leaves of some South American species ; the resin called dragon's-blood is the product of one of the rattan palms ; while the fruit of the Areca palm is the "betelnut" so universally chewed by the Malays as a gentle stimulant, and which is their substitute for the opium of the Chinese, the tobacco of Europeans, and the cocaleaf of South America.

For thatching, the leaves of palms are invaluable, and are universally used wherever they are abundant ; and the petioles or leaf-stalks, often fifteen or twenty feet long, are used as rafters, or when fastened together with pegs form doors, shutters, partitions, or even the walls of entire houses. They are wonderfully light and strong, being formed of a dense pith covered with a hard rind or bark, and when split up and pegged together serve to make many kinds of boxes, which, when covered with the broad leaves of a species of screw-pine and painted or stained of various colours, are very strong and serviceable as well as very ornamental. Ropes and cables are woven from the black fibrous matter that fringes the leaves of the sugar-palm and some other

species, while fine string of excellent quality used even
for bow-strings, fishing-lines, and hammocks, is made of
fibres obtained from the unopened leaves of some American
species.   The fibrous sheath at the base of the leaves of
the cocoa-nut palm is so compact and cloth-like, that it
is used for a variety of purposes, as for strainers, for
wrappers, and to make very good hats.   The great
woody spathes of the larger palms serve as natural
baskets, as cradles, or even as cooking-vessels in which
water may be safely boiled.   The trunks form excellent
posts and fencing, and when split make good flooring.
Some species are used for bows, others for blow-pipes;
the smaller species are sometimes used as needles or to
make fish-hooks, and the larger as arrows.   To describe
in detail all the uses to which palm-trees and their
products are applied in various parts of the world
might occupy a volume; but the preceding sketch will
serve to give an idea of how important a part is filled
by this noble family of plants, whether we regard them
as a portion of the beautiful vegetation of the tropics, or
in relation to the manners and customs, the lives and
the well-being of the indigenous inhabitants.

Ferns.—The type of plants which, next to palms,
most attracts attention in the equatorial zone, is perhaps
that of the ferns, which here display themselves in vast
profusion and variety.   They grow abundantly on rocks
and on decaying trees; they clothe the sides of ravines
and the margins of streams; they climb up the trees and
over bushes; they form tufts and hanging festoons
among the highest branches.   Some are as small as mosses,
others have huge fronds eight or ten feet long, while in
mountainous districts the most elegant of the group, the

tree-ferns, bear their graceful crowns on slender stems twenty to thirty, or even fifty feet high. It is this immense variety rather than any special features that characterises the fern-vegetation of the tropics. We have here almost every conceivable modification of size, form of fronds, position of spores, and habit of growth, in plants that still remain unmistakably ferns. Many climb over shrubs and bushes in a most elegant manner ; others cling closely to the bark of trees like ivy. The great birds'-nest fern (*Platycerium*) attaches its shell-like fronds high up on the trunks of lofty trees. Many small terrestrial species have digitate, or ovate, or ivy-shaped, or even whorled fronds, resembling at first sight those of some herbaceous flowering-plants. Their numbers may be judged from the fact that in the vicinity of Tarrapoto, in Peru, Dr. Spruce gathered 250 species of ferns, while the single volcanic mountains of Pangerango in Java (10,000 feet high) is said to have produced 300 species.

Ginger-worts and wild Bananas.—These plants, forming the families Zingiberaceæ and Musaceæ of botanists, are very conspicuous ornaments of the equatorial forests, on account of their large size, fine foliage, and handsome flowers. The bananas and plantains are well known as among the most luxuriant and beautiful productions of the tropics. Many species occur wild in the forests ; all have majestic foliage and handsome flowers, while some produce edible fruit. Of the ginger-worts (Zingiberaceæ and Marantaceæ), the well known cannas of our tropical gardens may be taken as representatives, but the equatorial species are very numerous and varied, often forming dense thickets in damp places, and adorning the

forest shades with their elegant and curious or showy
flowers. The maranths produce " arrow-root," while the
ginger-worts are highly aromatic, producing ginger,
cardamums, grains of paradise, turmeric and several
medicinal drugs. The Musaceæ produce the most valuable
of tropical fruits and foods. The banana is the variety
which is always eaten as a fruit, having a delicate
aromatic flavour ; the plantain is a larger variety which is
best cooked. Roasted in the green state it is an excellent
vegetable resembling roasted chestnuts ; when ripe it is
sometimes pulped and boiled with water, making a very
agreeable sweet soup ; or it is roasted, or cut into slices
and fried, in either form being a delicious tropical
substitute for fruit pudding. These plants are annuals,
producing one immense bunch of fruit. This bunch is
sometimes four or five feet long containing near
200 plantains, and often weighs about a hundred-
weight. They grow very close together, and Humboldt
calculated that an acre of plantains would supply more
food than could be obtained from the same extent of
ground by any other known plant. Well may it be said
that the plantain is the glory of the tropics, and well
was the species named by Linnæus—*Musa paradisiaca!*

Arums.—Another very characteristic and remarkable
group of tropical plants are the epiphytal and climbing
arums. These are known by their large, arrow-shaped,
dark green and glossy leaves, often curiously lobed or
incised, and sometimes reticulated with large open
spaces, as if pieces had been regularly eaten out of
them by some voracious insects. Sometimes they form
clusters of foliage on living or dead trees to which they
cling by their aerial roots. Others climb up the smooth

bark of large trees, sending out roots as they ascend which clasp around the trunk. Some mount straight up, others wind round the supporting trunks, and their large, handsome, and often highly-remarkable leaves, which spread out profusely all along the stem, render them one of the most striking forms of vegetation which adorn the damper and more luxuriant parts of the tropical forests of both hemispheres.

Screw-pines.—These singular plants, constituting the family Pandanaceæ of botanists, are very abundant in many parts of the Eastern tropics, while they are comparatively scarce in America. They somewhat resemble Yuccas, but have larger leaves which grow in a close spiral screw on the stem. Some are large and palm-like, and it is a curious sight to stand under these and look up at the huge vegetable screw formed by the bases of the long drooping leaves. Some have slender-branched trunks, which send out aerial roots ; others are stemless, consisting of an immense spiral cluster of stiff leaves ten or twelve feet long and only two or three inches wide. They abound most in sandy islands, while the larger species grow in swampy forests. Their large-clustered fruits, something like pineapples, are often of a red colour ; and their long stiff leaves are of great use for covering boxes and for many other domestic uses.

Orchids.—These interesting plants, so well known from the ardour with which they are cultivated on account of their beautiful and singular flowers, are pre-eminently tropical, and are probably more abundant in the mountains of the equatorial zone than in any other region. Here they are almost omnipresent in some of their countless forms. They grow on the stems, in the

E

forks or on the branches of trees ; they abound on fallen trunks ; they spread over rocks, or hang down the face of precipices ; while some, like our northern species, grow on the ground among grass and herbage. Some trees whose bark is especially well adapted for their support are crowded with them, and these form natural orchid-gardens. Some orchids are particularly fond of the decaying leaf-stalks of palms or of tree-ferns. Some grow best over water, others must be elevated on lofty trees and well exposed to sun and air. The wonderful variety in the form, structure, and colour of the flowers of orchids is well known ; but even our finest collections give an inadequate idea of the numbers of these plants that exist in the tropics, because a large proportion of them have quite inconspicuous flowers and are not worth cultivation. More than thirty years ago the number of known orchids was estimated by Dr. Lindley at 3,000 species, and it is not improbable that they may be now nearly doubled. But whatever may be the numbers of the collected and described orchids, those that still remain to be discovered must be enormous. Unlike ferns, the species have a very limited range, and it would require the systematic work of a good botanical collector during several years to exhaust any productive district—say such an island as Java—of its orchids. It is not there-fore at all improbable that this remarkable group may ultimately prove to be the most numerous in species of all the families of flowering plants.

Although there is a peculiarity of habit that enables one soon to detect an orchidaceous plant even when not in flower, yet they vary greatly in size and aspect. Some of the small creeping species are hardly larger

than mosses, while the large Grammatophyllums of
Borneo, which grow in the forks of trees, form a mass
of leafy stems ten feet long, and some of the terrestrial
species—as the American Sobralias—grow erect to an
equal height. The fleshy aerial roots of most species
give them a very peculiar aspect, as they often grow
to a great length in the open air, spread over the surface
of rocks, or attach themselves loosely to the bark of
trees, extracting nourishment from the rain and from
the aqueous vapour of the atmosphere. Yet notwithstanding the abundance and variety of orchids in the
equatorial forests they seldom produce much effect by
their flowers. This is due partly to the very large proportion of the species having quite inconspicuous flowers ;
and partly to the fact that the flowering season for each
kind lasts but a few weeks, while different species flower
almost every month in the year. It is also due to the
manner of growth of orchids, generally in single plants or
clumps which are seldom large or conspicuous as compared
with the great mass of vegetation around them. It is only
at long intervals that the traveller meets with anything
which recalls the splendour of our orchid-houses and
flower-shows. The slender-stalked golden Oncidiums of
the flooded forests of the Upper Amazon ; the grand
Cattleyas of the drier forests ; the Cælogynes of the
swamps, and the remarkable *Vanda lowii* of the hill
forests of Borneo, are the chief examples of orchid-
beauty that have impressed themselves on the memory
of the present writer during twelve years' wandering
in tropical forests. The last-named plant is unique
among orchids, its comparatively small cluster of leaves
sending out numerous flower-stems, which hang down

like cords to a length of eight feet, and are covered with numbers of large star-like crimson-spotted flowers.

Bamboos.—The gigantic grasses called bamboos can hardly be classed as typical plants of the tropical zone, because they appear to be absent from the entire African continent and are comparatively scarce in South America. They also extend beyond the geographical tropics in China and Japan as well as in Northern India.· It is however within the tropics and towards the equator that they attain their full size and beauty, and it is here that the species are most numerous and offer that variety of form, size, and quality, which renders them so admirable a boon to man. A fine clump of large bamboos is perhaps the most graceful of all vegetable forms, resembling the light and airy plumes of the bird-of-paradise copied on a gigantic scale in living foliage. Such clumps are often eighty or a hundred feet high, the glossy stems, perhaps six inches thick at the base, springing up at first straight as an arrow, tapering gradually to a slender point, and bending over in elegant curves with the weight of the slender branches and grassy leaves. The various species differ greatly in size and proportions; in the comparative length of the joints; in the thickness and strength of the stem-walls; in their straightness, smoothness, hardness, and durability. Some are spiny, others are unarmed; some have simple stems, others are thickly set with branches; while some species even grow in such an irregular, zig-zag, branched manner as to form veritable climbing bamboos. They generally prefer dry and upland stations, though some grow near the banks of rivers, and a few in the thick forests and, in South America, in flooded tracts. They often form dense thickets where the forests have been cleared away ; and, owing to their great utility, they are cultivated or preserved near native houses and villages, and in such situations often give a finishing charm to the landscape. *Uses of the Bamboo.*—Perhaps more than any other single type of vegetation, the bamboo seems specially adapted for the use of half-civilized man in a wild tropical country ; and the purposes to which it is applied are almost endless. It is a natural column or cylinder, very straight, uniform in thickness, of a compact and solid texture, and with a smooth flinty naturally-polished external skin. It is divided into ringed joints at regular intervals which correspond to *septa* or partitions within, so that each joint forms a perfectly closed and air-tight vessel. Owing to its hollowness, the hardness of the external skin, and the existence of the joints and partitions, it is wonderfully strong in proportion to its weight. It can be found of many distinct sizes and proportions ; light or heavy, long or short-jointed, and varying from the size of a reed to that of a tall and slender palm-tree. It can be split with great facility and accuracy ; and, owing to its being hollow, it can be easily cut across or notched with a sharp knife or hatchet. It is excessively strong and highly elastic, and whether green or dry is almost entirely free from any peculiar taste or smell. The way in which these various qualities of the bamboo render it so valuable, will be best shown by giving a brief account of some of the uses to which it is applied in the Malay Archipelago.

Several effective weapons are easily made from bamboo. By cutting off the end very obliquely just beyond a joint, a very sharp cutting point is produced suitable for a spear, dagger, or arrow-head, and capable of penetrating an animal's body as readily as iron. Such spears are constantly used by many of the Malay tribes. In the eastern half of the Archipelago, where bows and arrows are used, these weapons are often formed entirely of bamboo. The harder and thicker sorts, split and formed with tapering ends, make a very strong and elastic bow, while a narrow strip of the outer skin of the same is used for the string, and the slender reed-like kinds make excellent arrows. One of the few agricultural tools used by the Papuans—a spud or hoe for planting or weeding—is made of a stout bamboo cut somewhat like the spear.

For various domestic purposes the uses of bamboo are endless. Ladders are rapidly made from two bamboo poles of the required length, by cutting small notches just above each ring, forming holes to receive the rungs or steps formed of a slenderer bamboo. For climbing lofty trees to get beeswax, a temporary ladder reaching to any height is ingeniously formed of bamboo. One of the hardest and thickest sorts is chosen, and from this a number of pegs about a foot long are made. These are sharpened at one end and then driven into the tree in a vertical line about three feet apart. A tall and slender bamboo is then placed upright on the ground and securely tied with rattan or other cords to the heads of these pegs, which thus, with the tree itself, form a ladder. A man mounts these steps and builds up the ladder as he goes, driving in fresh pegs and splicing on fresh bamboos till he reaches the lower branches of the tree, which is sometimes eighty or a hundred feet from the ground. As the weight of the climber is thrown on several of the pegs, which are bound together and supported by the upright bamboo, this ladder is much safer that it looks at first sight, and it is made with wonderful rapidity. When a path goes up a steep hill over smooth ground, bamboo steps are often laid down to prevent slipping while carrying. heavy loads. These are made with uniform lengths of stout bamboo in which opposite notches are cut at each end just within a joint. These notches allow strong bamboo pegs to be driven through into the ground, thus keeping the steps securely in place. The masts and yards of native vessels are almost always formed of bamboo, as it combines lightness, strength, and elasticity in an unequalled degree. Two or three large bamboos also form the best outriggers to canoes on account of their great buoyancy. They also serve to form rafts; and in the city of Palembang in Sumatra there is a complete street of floating houses supported on rafts formed of huge bundles of bamboos. Bridges across streams or to carry footpaths along the face of precipices are constructed by the Dyaks of Borneo wholly of bamboos, and some of these are very ingeniously hung from overhanging trees by diagonal rods of bamboo, so as to form true suspension bridges. The flooring of Malay houses is almost always of bamboo, but is constructed in a variety of ways. Generally large bamboos are used, split lengthways twice and the pieces tied down with rattan. This forms a grated floor, slightly elastic, and very pleasant to the barefooted natives. A superior floor is sometimes formed of slabs, which are made from very stout bamboos cut into lengths of about three or four feet and split down one side. The joints are then deeply and closely notched all round with a sharp-chopping-knife, so that the piece can be unrolled as it were and pressed flat, when it forms a hard board with a natural surface which, with a little wear, becomes beautifully smooth and polished. Blinds, screens, and mats, are formed of bamboos in a variety of ways,—sometimes of thin kinds crushed flat and plaited, but more frequently of narrow strips connected together with cords of bamboo-bark or rattan. Strips of bamboo supported on cross-pieces form an excellent bed, which from its elasticity supplies the purpose of a mattress as well, and only requires a mat laid over it to insure a comfortable night's repose. Every kind of basket, too, is made of bamboo, from the coarsest heavy kinds to such as are fine and ornamental. In such countries as Lombock and Macassar, where the land is much cultivated and timber scarce, entire houses are built of bamboo,—posts, walls, floors, and roofs all being constructed of this one material ; and perhaps in no other way can so elegant and well-finished a house be built so quickly and so cheaply. Almost every kind of furniture is also made of the same material, excellent bamboo chairs, sofas, and bedsteads being made in the Moluccas, which, for appearance combined with cheapness, are probably unsurpassed in the world. A chair costs sixpence, and a sofa two shillings.

Among simpler uses, bamboos are admirably adapted for water-vessels. Some of the lighter sorts are cut into lengths of about five feet, a small hole being knocked through the septa of the joints. This prevents the water from running out too quickly, and facilitates its being poured out in a regulated stream to the last drop. Three or four of these water-vessels are tied together and carried on the back, and they stand very conveniently in a corner of the hut. Water pipes and aqueducts are also readily made from bamboo tubes supported at intervals on two smaller pieces tied crosswise. In this way a stream of water is often conveyed from some distance to the middle of a village. Measures for rice or palm-wine, drinking-vessels, and water-dippers, are to be found almost ready-made in a joint of bamboo ; and when fitted with a cap or lid they form tobacco or tinder-boxes. Perches for parrots with food and water vessels are easily made out of a single piece of bamboo, while with a little more labour elegant bird-cages are constructed. In Timor a musical instrument is formed from a single joint of a large bamboo, by carefully raising seven strips of the hard skin to form strings, which remain attached at both ends and are elevated by small pegs wedged underneath, the strings being prevented from splitting off by a strongly-plaited ring of a similar material bound round each end. An opening cut on one side allows the bamboo to vibrate in musical notes when the harp-like strings are sharply pulled with the fingers. In Java strips of bamboo supported on stretched strings and struck with a small stick produce the higher notes in the "gamelung" or native band, which consists mainly of sets of gongs and metallic plates of various sizes. Almost all the common Chinese paper is made from the foliage and stems of some species of bamboo, while the young shoots, as they first spring out of the ground, are an excellent vegetable, quite equal to artichokes. Single joints of bamboo make excellent cooking-vessels while on a journey. Rice can be boiled in them to perfection, as well as fish and vegetables. They serve too for jars in which to preserve sugar, salt, fruit, molasses, and cooked provisions ; and for the smoker, excellent pipes and hookahs can be formed in a few minutes out of properly chosen joints of bamboo.

These are only a sample of the endless purposes to which the bamboo is applied in the countries of which it is a native, its chief characteristic being that in a few minutes it can be put to uses which, if ordinary wood were used, would require hours or even days of labour. There is also a regularity and a finish about it which is found in hardly any other woody plant ; and its smooth and symmetrically ringed surface gives an appearance of fitness and beauty to its varied applications.   On the whole, we may perhaps consider it as the greatest boon which nature gives to the natives of the Eastern tropics.

Mangroves.—Among the forms of plants which are sure to attract attention in the tropics are the mangroves, which grow between tide-marks on coasts and estuaries. These are low trees with widely-spreading branches and a network of aerial roots a few feet above the ground ; but their most remarkable peculiarity is, that their fruits germinate on the tree, sending out roots and branches before falling into the muddy soil—a completely formed plant.   In some cases the root reaches the ground before the seed above falls off.   These trees greatly aid the formation of new land, as the mass of aerial roots which arch out from the stem to a considerable distance collects mud and floating refuse, and so raises and consolidates the shore ; while the young plants often dropping from the farthest extremity of the branches, rapidly extend the domain of vegetation to the farthest possible limits. The branches, too, send down slender roots like those of the banyan, and become independent trees. Thus a complete woody labyrinth is formed ; and the network of tough roots and stems resists the action of the tides, and enables the mud brought down by great tropical rivers to be converted into solid land far more rapidly than it could be without this aid.

Sensitive-plants.—Among the more humble forms of vegetation that attract the traveller's notice none are more interesting than the sensitive species of Mimosa. These are all natives of South America, but one species, *Mimosa pudica,* has spread to Africa and Asia, so that sensitive-plants now abound as wayside weeds in many parts both of the eastern and western tropics, sometimes completely carpeting the ground with their delicate foliage. Where a large surface of ground is thus covered the effect of walking over it is most peculiar. At each step the plants for some distance round suddenly droop, as if struck with paralysis, and a broad track of prostrate herbage, several feet wide, is distinctly marked out by the different colour of the closed leaflets. The explanation of this phenomenon, given by botanists, is not very satisfactory ; [1] while the purpose or use of the peculiarity is still more mysterious, seeing that out of about two hundred species belonging to this same genus Mimosa, only some three or four are sensitive, and in the whole vegetable kingdom there are no other plants which possess more than the rudiments of a similar property.

[1] See *Nature,* vol. xvi. p. 349, where the German botanist Pfeffer's theory is given.

It is true that, as they are all low-growing herbs or shrubs with delicate foliage, they might possibly be liable to destruction by herbivorous animals, and might escape by their singular power of suddenly collapsing before the jaws opened to devour them. The fact that one species has been naturalized as a weed over so wide an area in the tropics, seems to show that it possesses some advantage over the generality of tropical weeds. It is however curious that, as most of the species are somewhat prickly, so easy and common a mode of protection as the development of stronger spines should here have failed; and that its place should be supplied by so singular a power as that of simulating death, in a manner which suggests the possession of both sensation and voluntary motion.

Comparative Scarcity of Flowers.—It is a very general opinion among inhabitants of our temperate climes, that amid the luxuriant vegetation of the tropics there must be a grand display of floral beauty; and this idea is supported by the number of large and showy flowers cultivated in our hot-houses. The fact is, however, that in proportion as the general vegetation becomes more luxuriant, flowers form a less and less prominent feature; and this rule applies not only to the tropics but to the temperate and frigid zones. It is amid the scanty vegetation of the higher mountains and towards the limits of perpetual snow, that the alpine flowers are most brilliant and conspicuous. Our own meadows and pastures and hill-sides produce more gay flowers than our woods and forests; and, in the tropics, it is in the parts where vegetation is less dense and luxuriant that flowers most abound. In the damp and uniform climate of the equatorial zone the mass of vegetation is greater and more varied than in any other part of the globe, but in the great virgin forests themselves flowers are rarely seen. After describing the forests of the Lower Amazon, Mr. Bates asks : "But where were the flowers ? To our great disappointment we saw none, or only such as were insignificant in appearance. Orchids are rare in the dense forests of the lowlands, and I believe it is now tolerably well ascertained that the majority of the forest-trees in equatorial Brazil have small and inconspicuous flowers." [1] My friend Dr. Richard Spruce assured me that by far the greater part of the plants gathered by him in equatorial America had inconspicuous green or white flowers. My own observations in the Aru Islands for six months, and in Borneo for more than a year, while living almost wholly in the forests, are quite in accordance with this view. Conspicuous masses of showy flowers are so rare, that weeks and months may be passed without observing a single flowering plant worthy of special admiration. Occasionally some tree or shrub will be seen covered with magnificent yellow, or crimson, or purple flowers, but it is usually an oasis of colour in a desert of verdure, and therefore hardly affects the general aspect of the vegetation. The equatorial forest is too gloomy for flowers, or generally even for much foliage, except of ferns and other shade-loving plants ; and were it not that the forests are broken up by rivers and streams, by mountain ranges, by precipitous rocks and by deep ravines, there would be far fewer flowers than there are. Some of the great forest-trees have showy blossoms,

[1] *The Naturalist on the River Amazons*, 2nd edit. p. 38.

and when these are seen from an elevated point looking over an expanse of tree-tops the effect is very grand ; but nothing is more erroneous than the statement sometimes made that tropical forest-trees *generally* have showy flowers, for it is doubtful whether the proportion is at all greater in tropical than in temperate zones. On such natural exposures as steep mountain sides, the banks of rivers, or ledges of precipices, and on the margins of such artificial openings as roads and forest clearings, whatever floral beauty is to be found in the more luxuriant parts of the tropics is exhibited. But even in such favourable situations it is not the abundance and beauty of the flowers but the luxuriance and the freshness of the foliage, and the grace and infinite variety of the forms of vegetation, that will most attract the attention and extort the admiration of the traveller. Occasionally indeed you will come upon shrubs gay with blossoms or trees festooned with flowering creepers ; but, on the other hand, you may travel for a hundred miles and see nothing but the varied greens of the forest foliage and the deep gloom of its tangled recesses. In Mr. Belt's *Naturalist in Nicaragua,* he thus describes the great virgin forests of that country which, being in a mountainous region and on the margin of the equatorial zone, are among the most favourable examples. " On each side of the road great trees towered up, carrying their crowns out of sight amongst a canopy of foliage, and with lianas hanging from nearly every bough, and passing from tree to tree, entangling the giants in a great network of coiling cables. Sometimes a tree appears covered with beautiful flowers which do not belong to it, but to one of the lianas that twines through its branches and sends down great rope-like stems to the ground. Climbing ferns and vanilla cling to the trunks, and a thousand epiphytes perch themselves on the branches. Amongst these are large arums that send down long aerial roots, tough and strong, and universally used instead of cordage by the natives. Amongst the undergrowth several small species of palms, varying in height from two to fifteen feet, are common ; and now and then magnificent tree ferns sending off their feathery crowns twenty feet from the ground delight the sight by their graceful elegance. Great broad-leaved heliconias, leathery melastomæ, and succulent-stemmed, lop-sided leaved and flesh-coloured begonias are abundant, and typical of tropical American forests ; but not less so are the cecropia trees, with their white stems and large palmated leaves standing up like great candelabra. Sometimes the ground is carpeted with large flowers, yellow, pink, or white, that have fallen from some invisible tree-top above ; or the air is filled with a delicious perfume, the source of which one seeks around in vain, for the flowers that cause it are far overhead out of sight, lost in the great overshadowing crown of verdure."

Although, as has been shown elsewhere, it may be doubted whether light directly produces floral colour, there can be no doubt that it is essential to the growth of vegetation and to the full development of foliage and of flowers. In the forests all trees, and shrubs, and creepers struggle upwards to the light, there to expand their blossoms and ripen their fruit. Hence, perhaps, the abundance of climbers which make use of their more sturdy companions to reach this necessary of vegetable

life. Yet even on the upper surface of the forest, fully
exposed to the light and heat of the tropical sun,
there is no special development of coloured flowers.
When from some elevated point you can gaze down upon
an unbroken expanse of woody vegetation, it often
happens that not a single patch of bright colour can be
discerned. At other times, and especially at the
beginning of the dry season, you may behold scattered
at wide intervals over the mottled-green surface a few
masses of yellow, white, pink, or more rarely of blue
colour, indicating the position of handsome flowering
trees.

The well-established relation between coloured flowers
and the need of insects to fertilize them, may perhaps be
connected with the comparative scarcity of the former
in the equatorial forests. The various forms of life are
linked together in such mutual dependence that no one
can inordinately increase without bringing about a
corresponding increase or diminution of other forms.
The insects which are best adapted to fertilize flowers
cannot probably increase much beyond definite limits,
because in doing so they would lead to a corresponding
increase of insectivorous birds and other animals which
would keep them down. The chief fertilizers—bees and
butterflies—have enemies at every stage of their growth,
from the egg to the perfect insect, and their numbers are,
therefore, limited by causes quite independent of the
supply of vegetable food. It may, therefore, be the case
that the numbers of suitable insects are totally inadequate to the fertilization of the countless millions of
forest-trees over such vast areas as the equatorial zone
presents, and that, in consequence, a large proportion of

the species have become adapted either for self-fertilization or for cross-fertilization by the agency of the wind.
Were there not some such limitation as this, we should
expect that the continued struggle for existence among
the plants of the tropical forests would have led to the
acquisition, by a much larger proportion of them, of so
valuable a character as bright-coloured flowers, this being
almost a necessary preliminary to a participation in the
benefits which have been proved to arise from cross-
fertilization by insect agency.

Concluding Remarks on Tropical Vegetation.—In
concluding this general sketch of the aspect of tropical
vegetation we will attempt briefly to summarize its main
features.   The primeval forests of the equatorial zone are
grand and overwhelming by their vastness, and by the
display of a force of development and vigour of growth
rarely or never witnessed in temperate climates.   Among
their best distinguishing features are the variety of forms
and species which everywhere meet and grow side by side,
and the extent to which parasites, epiphytes, and creepers
fill up every available station with peculiar modes of life.
If the traveller notices a particular species and wishes to
find more like it, he may often turn his eyes in vain in
every direction.   Trees of varied forms, dimensions, and
colours are around him, but he rarely sees any one of
them repeated.   Time after time he goes towards a tree
which looks like the one he seeks, but a closer examination proves it to be distinct.   He may at length,
perhaps, meet with a second specimen half a mile off, or
may fail altogether, till on another occasion he stumbles
on one by accident.

The absence of the gregarious or social habit, so

general in the forests of extra-tropical countries, is probably dependent on the extreme equability and permanence of the climate. Atmospheric conditions are much more important to the growth of plants than any others. Their severest struggle for existence is against climate. As we approach towards regions of polar cold or desert aridity the variety of groups and species regularly diminishes ; more and more are unable to sustain the extreme climatal conditions, till at last we find only a few specially organized forms which are able to maintain their existence. In the extreme north, pine or birch trees ; in the desert, a few palms and prickly shrubs or aromatic herbs alone survive. In the equable equatorial zone there is no such struggle against climate. Every form of vegetation has become alike adapted to its genial heat and ample moisture, which has probably changed little even throughout geological periods ; and the never-ceasing struggle for existence between the various species in the same area has resulted in a nice balance of organic forces, which gives the advantage, now to one, now to another, species, and prevents any one type of vegetation from monopolising territory to the exclusion of the rest. The same general causes have led to the filling up of every place in nature with some specially adapted form. Thus we find a forest of smaller trees adapted to grow in the shade of greater trees. Thus we find every tree supporting numerous other forms of vegetation, and some so crowded with epiphytes of various kinds that their forks and horizontal branches are veritable gardens. Creeping ferns and arums run up the smoothest trunks ; an immense variety of climbers hang in tangled masses from the branches and mount over the highest tree-tops. Orchids, bromelias, arums, and ferns grow from every boss and crevice, and cover the fallen and decaying trunks with a graceful drapery. Even these parasites have their own parasitical growth, their leaves often supporting an abundance of minute creeping mosses and hepaticæ. But the uniformity of climate which has led to this rich luxuriance and endless variety of vegetation is also the cause of a monotony that in time becomes oppressive. To quote the words of Mr. Belt : " Unknown are the autumn tints, the bright browns and yellows of English woods ; much less the crimsons, purples, and yellows of Canada, where the dying foliage rivals, nay, excels, the expiring dolphin in splendour. Unknown the cold sleep of winter ; unknown the lovely awakening of vegetation at the first gentle touch of spring. A ceaseless round of ever-active life weaves the fairest scenery of the tropics into one monotonous whole, of which the component parts exhibit in detail untold variety and beauty." [1]

To the student of nature the vegetation of the tropics will ever be of surpassing interest, whether for the variety of forms and structures which it presents, for the boundless energy with which the life of plants is therein manifested, or for the help which it gives us in our search after the laws which have determined the production of such infinitely varied organisms. When, for the first time, the traveller wanders in these primeval forests, he can scarcely fail to experience sensations of awe, akin to those excited by the trackless ocean or the alpine snowfields. There is a vastness, a solemnity, a gloom, a sense of solitude and of human insignificance

[1] *The Naturalist in Nicaragua*, p. 58.

which for a time overwhelm him ; and it is only when the novelty of these feelings have passed away that he is able to turn his attention to the separate constituents that combine to produce these emotions, and examine the varied and beautiful forms of life which, in inexhaustible profusion, are spread around him.

III.

ANIMAL LIFE IN THE TROPICAL FORESTS.

Difficulties of the Subject—General Aspect of the Animal life of Equatorial Forests—Diurnal Lepidoptera or Butterflies—Peculiar Habits of Tropical Butterflies—Ants, Wasps, and Bees—Ants—Special Relations between Ants and Vegetation—Wasps and Bees—Orthoptera and other Insects—Beetles—Wingless Insects—General Observations on Tropical Insects—Birds—Parrots—Pigeons—Picariæ—Cuckoos—Trogons, Barbets, Toucans and Hornbills—Passeres—Reptiles and Amphibia—Lizards—Snakes—Frogs and Toads — Mammalia — Monkeys — Bats — Summary of the Aspects of Animal life in the Tropics.

THE attempt to give some account of the general aspects of animal life in the equatorial zone, presents far greater difficulties than in the case of plants. On the one hand, animals rarely play any important part in scenery, and their entire absence may pass quite unnoticed ; while the abundance, variety, and character of the vegetation are among those essential features that attract every eye. On the other hand, so many of the more important and characteristic types of animal life are restricted to one only out of the three great divisions of equatorial land, that they can hardly be claimed as characteristically tropical ; while the more extensive zoological groups which have a wide range in the tropics and do not equally abound in the temperate zones, are few in number, and often include such a diversity of forms, structures, and habits, as to render any typical characterisation of them impossible. We must then, in the first place, suppose that our traveller is on the look out for all signs of animal life ; and that, possessing a general acquaintance as an out-door observer with the animals of our own country, he carefully notes those points in which the forests of the equatorial zone offer different phenomena. Here, as in the case of plants, we exclude all zoological science, classifications, and nomenclature, except in as far as it is necessary for a clear understanding of the several groups of animals referred to. We shall therefore follow no systematic order in our notes, except that which would naturally arise from the abundance or prominence of the objects themselves. We further suppose our traveller to have no prepossessions, and to have no favourite group, in the search after which he passes by other objects which, in view of their frequent occurrence in the landscape, are really more important.

General Aspect of the Animal Life of Equatorial Forests.—Perhaps the most general impression produced by a first acquaintance with the equatorial forests, is the comparative absence of animal life. Beast, bird, and insect alike require looking for, and it very often happens that we look for them in vain. On this subject Mr. Bates, describing one of his early excursions into the primeval forests of the Amazon Valley, remarks as follows :—" We were disappointed in not meeting with any of the larger animals of the forest. There was no tumultuous movement or sound of life. We did not see or hear monkeys, and no tapir or jaguar crossed our path. Birds also appeared to be exceedingly scarce."

Again—" I afterwards saw reason to modify my opinion, founded on first impressions, with regard to the amount and variety of animal life in this and other parts of the Amazonian forests. There is in fact a great variety of mammals, birds, and reptiles, but they are widely scattered and all excessively shy of man. The region is so extensive, and uniform in the forest clothing of its surface, that it is only at long intervals that animals are seen in abundance, where some particular spot is found which is more attractive than others. Brazil, moreover, is throughout poor in terrestrial mammals, and the species are of small size ; they do not, therefore, form a conspicuous feature in the forests. The huntsman would be disappointed who expected to find here flocks of animals similar to the buffalo-herds of North America, or the swarms of antelopes and herds of ponderous pachyderms of Southern Africa. We often read in books of travel of the silence and gloom of the Brazilian forests. They are realities, and the impression deepens on a longer acquaintance. The few sounds of birds are of that pensive and mysterious character which intensifies the feeling of solitude rather than imparts a sense of life and cheerfulness. Sometimes in the midst of the stillness, a sudden yell or scream will startle one ; this comes from some defenceless fruit-eating animal which is pounced upon by a tiger-cat or a boa-constrictor. Morning and evening the howling monkeys make a most fearful and harrowing noise, under which it is difficult to keep up one's buoyancy of spirit. The feeling of inhospitable wildness which the forest is calculated to inspire, is increased tenfold under this fearful uproar. Often, even in the still midday hours, a sudden crash will be heard

resounding afar through the wilderness, as some great
bough or entire tree falls to the ground." With a few
verbal alterations these remarks will apply equally to
the primeval forests of the Malay Archipelago ; and it is
probable that those of West Africa offer no important
differences in this respect. There is, nevertheless, one
form of life which is very rarely absent in the more
luxuriant parts of the tropics, and which is more often
so abundant as to form a decided feature in the scene.
It is therefore the group which best characterises the
equatorial zone, and should form the starting-point for
our review. This group is that of the diurnal Lepidóp-
tera or butterflies.

Diurnal Lepidoptera.—Wherever in the equatorial
zone a considerable extent of the primeval forest
remains, the observer can hardly fail to be struck by the
abundance and the conspicuous beauty of the butterflies.
Not only are they abundant in individuals, but their
large size, their elegant forms, their rich and varied
colours, and the number of distinct species almost
everywhere to be met with are equally remarkable. In
many localities near the northern or southern tropics
they are perhaps equally abundant, but these spots are
more or less exceptional ; whereas within the equatorial
zone, and with the limitations above stated, butterflies
form one of the most constant and most conspicuous
displays of animal life. They abound most in old and
tolerably open roads and pathways through the forest,
but they are also very plentiful in old settlements in
which fruit-trees and shrubbery offer suitable haunts. In
the vicinity of such old towns as Malacca and Amboyna
in the East, and of Para and Rio de Janeiro in the

West, they are especially abundant, and comprise some
of the handsomest and most remarkable species in the
whole group. Their aspect is altogether different from
that presented by the butterflies of Europe and of most
temperate countries. A considerable proportion of the
species are very large, six to eight inches across the
wings being not uncommon among the Papilionidæ and
Morphidæ, while several species are even larger. This
great expanse of wings is accompanied by a slow flight ;
and, as they usually keep near the ground and often
rest, sometimes with closed and sometimes with expanded wings, these noble insects really look larger and
are much more conspicuous objects than the majority of
our native birds. The first sight of the great blue
Morphos flapping slowly along in the forest roads near
Para—of the large, white-and-black semi-transparent
Ideas floating airily about in the woods near Malacca—
and of the golden-green Ornithopteras sailing on bird-
like wing over the flowering shrubs which adorn the
beach of the Ké and Aru islands, can never be forgotten
by any one with a feeling of admiration for the new and
beautiful in nature. Next to the size, the infinitely
varied and dazzling hues of these insects most attract
the observer. Instead of the sober browns, the plain
yellows, and the occasional patches of red or blue or
orange that adorn our European species, we meet with
the most intense metallic blues, the purest satiny greens,
the most gorgeous crimsons, not in small spots but in
large masses, relieved by a black border or background.
In others we have contrasted bands of blue and orange,
or of crimson and green, or of silky yellow relieved by
velvety black. In not a few the wings are powdered

over with scales and spangles of metallic green, deepening occasionally into blue or golden or deep red spots. Others again have spots and markings as of molten silver or gold, while several have changeable hues, like shot-silk or richly-coloured opal. The form of the wings, again, often attracts attention. Tailed hind-wings occur in almost all the families, but vary much in character. In some the tails are broadly spoon-shaped, in others long and pointed. Many have double or triple tails, and some of the smaller species have them immensely elongated and often elegantly curled. In some groups the wings are long and narrow, in others strongly falcate ; and though many fly with immense rapidity, a large number flutter lazily along, as if they had no enemies to fear and therefore no occasion to hurry.

The number of species of butterflies inhabiting any one locality is very variable, and is, as a rule, far larger in America than in the Eastern hemisphere ; but it everywhere very much surpasses the numbers in the temperate zone. A few months' assiduous collecting in any of the Malay islands will produce from 150 to 250 species of butterflies, and thirty or forty species may be obtained any fine day in good localities. In the Amazon valley, however, much greater results may be achieved. A good day's collecting will produce from forty to seventy species, while in one year at Para about 600 species were obtained. More than 700 species of butterflies actually inhabit the district immediately around the city of Para, and this, as far as we yet know, is the richest spot on the globe for diurnal lepidoptera. At Ega, during four years' collecting ;

Mr. Bates obtained 550 species, and these on the whole surpassed those of Para in variety and beauty. Mr. Bates thus speaks of a favourite locality on the margin of the lake near Ega :—" The number and variety of gaily-tinted butterflies, sporting about in this grove on sunny days, were so great, that the bright moving flakes of colour gave quite a character to the physiognomy of the place. It was impossible to walk far without disturbing flocks of them from the damp sand at the edge of the water, where they congregated to imbibe the moisture. They were of almost all colours, sizes, and shapes ; I noticed here altogether eighty species, belonging to twenty-two distinct genera. The most abundant, next to the very common sulphur-yellow and orange-coloured kinds, were about a dozen species of Eunica, which are of large size and conspicuous from their liveries of glossy dark blue and purple. A superbly adorned creature, the Callithea Markii, having wings of a thick texture, coloured sapphire-blue and orange, was only an occasional visitor. On certain days, when the weather was very calm, two small gilded species (Symmachia Trochilus and Colubris) literally swarmed on the sands, their glittering wings lying wide open on the flat surface." [1]

When we consider that only sixty-four species of butterflies have been found in Britain and about 150 in Germany, many of which are very rare and local, so that these numbers are the result of the work of hundreds of collectors for a long series of years, we see at once the immense wealth of the equatorial zone in this form of life.

[1] *The Naturalist on the Amazons*, 2nd edit. p. 331.

Peculiar Habits of Tropical Butterflies.—The habits of the butterflies of the tropics offer many curious points rarely or never observed among those of the temperate zone. The majority, as with us, are truly diurnal, but there are some Eastern Morphidæ and the entire American family Brassolidæ, which are crepuscular, coming out after sunset and flitting about the roads till it is nearly dark. Others, though flying in the daytime, are only found in the gloomiest recesses of the forest, where a constant twilight may be said to prevail. The majority of the species fly at a moderate height (from five to ten feet above the ground) while a few usually keep higher up and are difficult to capture ; but a large number, especially the Satyridæ, many Erycinidæ, and some few Nymphalidæ, keep always close to the ground, and usually settle on or among the lowest herbage. As regards the mode of flight, the extensive and almost exclusively tropical families of Heliconidæ and Danaidæ, fly very slowly, with a gentle undulating or floating motion which is almost peculiar to them. Many of the strong-bodied Nymphalidæ and Hesperidæ, on the other hand, have an excessively rapid flight, darting by so swiftly that the eye cannot follow them, and in some cases producing a deep sound louder than that of the humming-birds.

The places they frequent, and their mode of resting, are various and often remarkable. A considerable number frequent damp open places, especially river sides and the margins of pools, assembling together in flocks of hundreds of individuals ; but these are almost entirely composed of males, the females remaining in the forests where, towards the afternoon, their partners join them.

The majority of butterflies settle upon foliage and on flowers, holding their wings erect and folded together, though early in the morning, or when newly emerged from the chrysalis, they often expand them to the sun. Many, however, have special stations and attitudes. Some settle always on tree-trunks, usually with the wings erect, but the Ageronias expand them and always rest with the head downwards. Many Nymphalidæ prefer resting on the top of a stick ; others choose bushes with dead leaves ; others settle on rocks or sand or in dry forest paths. Pieces of decaying animal or vegetable matter are very attractive to certain species, and if disturbed they will sometimes return to the same spot day after day. Some Hesperidæ, as well as species of the genera Cyrestis and Symmachia, and some others, rest on the ground with their wings fully expanded and pressed closely to the surface, as if exhibiting themselves to the greatest advantage. The beautiful little Erycinidæ of South America vary remarkably in their mode of resting. The majority always rest on the under surface of leaves with their wings expanded, so that when they settle they suddenly disappear from sight. Some, however, as the elegant gold-spotted Helicopis cupido, rest beneath leaves with closed wings. A few, as the genera Charis and Themone, for example, sit on the upper side of leaves with their wings expanded ; while the gorgeously-coloured Erycinas rest with wings erect and exposed as in the majority of butterflies. The Hesperidæ vary in a somewhat similar manner. All rest on the upper side of leaves or on the ground, but some close their wings, others expand them, and a third group keep the upper pair of wings raised while the hind wings are expanded, a habit found in some of our European species. Many of the Lycænidæ, especially the Theclas, have the curious habit, while sitting with their wings erect, of moving the lower pair over each other in opposite directions, giving them the strange appearance of excentrically revolving discs.

The great majority of butterflies disappear at night, resting concealed amid foliage, or on sticks or trunks, or in such places as harmonise with their colours and markings ; but the gaily-coloured Heliconidæ and Danaidæ seek no such concealment, but rest at night hanging at the ends of slender twigs or upon fully exposed leaves. Being uneatable they have no enemies and need no concealment. Day-flying moths of brilliant or conspicuous colours are also comparatively abundant in the tropical forests. Most magnificent of all are the Uranias, whose long-tailed green-and-gold powdered wings resemble those of true swallow-tailed butterflies. Many Agaristidæ of the East are hardly inferior in splendour, while hosts of beautiful clear-wings and Ægeriidæ add greatly to the insect beauty of the equatorial zone.

The wonderful examples afforded by tropical butterflies of the phenomena of sexual and local variation, of protective modifications, and of mimicry, have been fully discussed elsewhere. For the study of the laws of variation in all its forms, these beautiful creatures are unsurpassed by any class of animals ; both on account of their great abundance, and the assiduity with which they have been collected and studied. Perhaps no group exhibits the distinctions of species and genera with such precision and distinctness, due, as Mr. Bates has well observed, to the fact that all the superficial signs of change in the organization are exaggerated, by their affecting the size, shape, and colour, of the wings, and the distribution of the ribs or veins which form their framework. The minute scales or feathers with which the wings are clothed are coloured in regular patterns, which vary in accordance with the slightest change in the conditions to which the species are exposed. These scales are sometimes absent in spots or patches, and sometimes over the greater part of the wings, which then become transparent, relieved only by the dark veins and by delicate shades or small spots of vivid colour, producing a special form of delicate beauty characteristic of many South American butterflies. The following remark by Mr. Bates will fitly conclude our sketch of these lovely insects :—" It may be said, therefore, that on these expanded membranes Nature writes, as on a tablet, the story of the modifications of species, so truly do all the changes of the organization register themselves thereon. And as the laws of Nature must be the same for all beings, the conclusions furnished by this group of insects must be applicable to the whole organic world ; therefore the study of butterflies—creatures selected as the types of airiness and frivolity—instead of being despised, will some day be valued as one of the most important branches of biological science." [1]

Next after the butterflies in importance, as giving an air of life and interest to tropical nature, we must place the birds ; but to avoid unnecessary passage, to and fro, among unrelated groups, it will be best to follow on

---

[1] Bates, *The Naturalist on the Amazons*, 2nd edit. p. 413.

with a sketch of such other groups of insects as from their numbers, variety, habits, or other important features, attract the attention of the traveller from colder climates. We begin then with a group, which owing to their small size and obscure colours would attract little attention, but which nevertheless, by the universality of their presence, their curious habits, and the annoyance they often cause to man, are sure to force themselves upon the attention of every one who visits the tropics.

Ants, Wasps, and Bees.—The hymenopterous insects of the tropics are, next to the butterflies, those which come most prominently before the traveller, as they love the sunshine, frequent gardens, houses, and roadways as well as the forest shades, never seek concealment, and are many of them remarkable for their size or form, or are adorned with beautiful colours and conspicuous markings. Although ants are, perhaps, on the whole the smallest and the least attractive in appearance of all tropical insects, yet, owing to their being excessively abundant and almost omnipresent, as well as on account of their curious habits and the necessity of being ever on the watch against their destructive powers, they deserve our first notice.

Ants are found everywhere. They abound in houses, some living underground, others in the thatched roof on the under surface of which they make their nests, while covered ways of earth are often constructed upon the posts and doors. In the forests they live on the ground, under leaves, on the branches of trees, or under rotten bark ; while others actually dwell in living plants, which seem to be specially modified so as to accommodate them. Some sting severely, others only bite ; some are quite

harmless, others exceedingly destructive. The number
of different kinds is very great. In India and the
Malay Archipelago nearly 500 different species have been
found, and other tropical countries are no doubt equally
rich. I will first give some account of the various
species observed in the Malay Islands, and afterwards
describe some of the more interesting South American
groups, which have been so carefully observed by Mr
Bates on the Amazon and by Mr. Belt in Nicaragua.

Among the very commonest ants in all parts of the
world are the species of the family Formicidæ, which do
not sting, and are most of them quite harmless. Some
make delicate papery nests, others live under stones or
among grass. Several of them accompany Aphides to
feed upon the sweet secretions from their bodies. They
vary in size from the large *Formica gigas*, more than an
inch long, to minute species so small as to be hardly
visible. Those of the genus Polyrachis, which are
plentiful in all Eastern forests, are remarkable for the
extraordinary hooks and spines with which their bodies
are armed, and they are also in many cases beautifully
sculptured or furrowed. They are not numerous individually, and are almost all arboreal, crawling about
bark and foliage. One species has processes on its
back just like fish-hooks, others are armed with long,
straight spines. They generally form papery nests on
leaves, and when disturbed they rush out and strike
their bodies against the nest so as to produce a loud
rattling noise ; but the nest of every species differs from
those of all others either in size, shape, or position. As
they all live in rather small communities in exposed
situations, are not very active, and are rather large and

conspicuous, they must be very much exposed to the attacks of insectivorous birds and other creatures ; and, having no sting or powerful jaws with which to defend themselves, they would be liable to extermination without some special protection. This protection they no doubt obtain by their hard smooth bodies, and by the curious hooks, spines, points and bristles with which they are armed, which must render them unpalatable morsels, very liable to stick in the jaws or throats of their captors.

A curious and very common species in the Malay Islands is the green ant (*Œcophylla smaragdina*), a rather large, long-legged, active, and intelligent-looking creature, which lives in large nests formed by glueing together the edges of leaves, especially of Zingiberaceous plants. When the nest is touched a number of the ants rush out, apparently in a great rage, stand erect, and make a loud rattling noise by tapping against the leaves. This no doubt frightens away many enemies, and is their only protection ; for though they attempt to bite, their jaws are blunt and feeble, and they do not cause any pain.

Coming now to the stinging groups, we have first a number of solitary ants of the great genus Odontomachus, which are seen wandering about the forest, and are conspicuous by their enormously long and slender hooked jaws. These are not powerful, but serve admirably to hold on by while they sting, which they do pretty severely. The Poneridæ are another group of large-sized ants which sting acutely. They are very varied in species but are not abundant individually. The *Ponera clavata* of Guiana, is one of the worst stinging ants known. It is a large species frequenting the forests on the ground, and is much dreaded by the natives, as its sting produces intense pain and illness. I was myself stung by this or an allied species when walking barefoot in the forest on the Upper Rio Negro. It caused such pain and swelling of the leg that I had some difficulty in reaching home, and was confined to my room for two days. Sir Robert Schomburgh suffered more ; for he fainted with the pain, and had an attack of fever in consequence.

We now come to the Myrmecidæ, which may be called the destroying ants from their immense abundance and destructive propensities. Many of them sting most acutely, causing a pain like that of a sudden burn, whence they are often called "fire-ants." They often swarm in houses and devour everything eatable. Isolation by water is the only security, and even this does not always succeed, as a little dust on the surface will enable the smaller species to get across. Oil is, however, an effectual protection, and after many losses of valuable insect specimens, for which ants have a special affection, I always used it. One species of this group, a small black Crematogaster, took possession of my house in New Guinea, building nests in the roof and making covered ways down the posts and across the floor. They also occupied the setting boards I used for pinning out my butterflies, filling up the grooves with cells and storing them with small spiders. They were in constant motion, running over my table, in my bed, and all over my body. Luckily, they were diurnal, so that on sweeping out my bed at night I could get on pretty well ; but during the day I could always feel some of them running over my body, and every now and then one would give me a sting so sharp as to make me jump and search instantly for the offender, who was usually found holding on tight with his jaws, and thrusting in his sting with all his might. Another genus, Pheidole, consists of forest ants, living under rotten bark or in the ground, and very voracious. They are brown or blackish, and are remarkable for their great variety of size and form in the same species, the largest having enormous heads many times larger than their bodies, and being at least a hundred times as bulky as the smallest individuals. These great-headed ants are very sluggish and incapable of keeping up with the more active small workers, which often surround and drag them along as if they were wounded soldiers. It is difficult to see what use they can be in the colony, unless, as Mr. Bates suggests, they are mere baits to be attacked by insect-eating birds, and thus save their more useful companions. These ants devour grubs, white ants, and other soft and helpless insects, and seem to take the place of the foraging ants of America and driver-ants of Africa, though they are far less numerous and less destructive. An allied genus, Solenopsis, consists of red ants, which, in the Moluccas, frequent houses, and are a most terrible pest. They form colonies underground, and work their way up through the floors, devouring everything eatable. Their sting is excessively painful, and some of the species are hence called fire-ants. When a house is infested by them, all the tables and boxes must be supported on blocks of wood or stone placed in dishes of water, as even clothes not newly washed are attractive to them ; and woe to the poor fellow who puts on garments in the folds of which a dozen of these ants are lodged. It is very difficult to preserve bird skins or other specimens of natural history where these ants abound, as they gnaw away the skin round the eyes and the base of the bill; and if a specimen is laid down for even half an hour in an unprotected place it will be ruined. I remember once entering a native house to rest and eat my lunch; and having a large tin collecting box full of rare butterflies and other insects, I laid it down on the bench by my side. On leaving the house I noticed some ants on it, and on opening the box found only a mass of detached wings and bodies, the latter in process of being devoured by hundreds of fire-ants.

The celebrated Saüba ant of America (*Œcodoma cephalotes*) is allied to the preceding, but is even more destructive, though it seems to confine itself to vegetable products. It forms extensive underground galleries, and the earth brought up is deposited on the surface, forming huge mounds sometimes thirty or forty yards in circumference, and from one to three feet high. On first seeing these vast deposits of red or yellow earth in the woods near Para, it was hardly possible to believe they were not the work of man, or at least of some burrowing animal. In these underground caves the ants store up large quantities of leaves, which they obtain from living trees. They gnaw out circular pieces and carry them away along regular paths a few inches wide, forming a stream of apparently animated leaves. The great extent of the subterranean workings of these ants is no doubt due in part to their permanence in one spot, so that when portions of the galleries fall in or are otherwise rendered useless, they are extended in another direction. When in the island of Marajo, near Para, I noticed a path along which a stream of Saübas were carrying leaves from a neighbouring thicket; and a relation of the proprietor assured me that he had known that identical path to be in constant use by the ants for twenty years. Thus we can account for the fact mentioned by Mr. Bates, that the underground galleries were traced by smoke for a distance of seventy yards in the Botanic Gardens at Para ; and for the still more extraordinary fact related by the Rev. Hamlet Clark, that an allied species in Rio de Janeiro has excavated a tunnel under the bed of the river Parahyba, where it is about a quarter of a mile wide ! These ants seem to prefer introduced to native trees ; and young plantations of orange, coffee, or mango trees are sometimes destroyed by them, so that where they abound cultivation of any kind becomes almost impossible. Mr. Belt ingeniously accounts for this preference, by supposing that for ages there has been a kind of struggle going on between the trees and the ants ; those varieties of trees which were in any way distasteful or unsuitable escaping destruction, while the ants were becoming slowly adapted to attack new trees. Thus in time the great majority of native trees have acquired some protection against the ants, while foreign trees, not having been so modified, are more likely to be suitable for their purposes. Mr. Belt carried on war against them for four years to protect his garden in Nicaragua, and found that carbolic acid and corrosive sublimate were most effectual in destroying or driving them away.

The use to which the ants put the immense quantities of leaves they carry away has been a great puzzle, and

is, perhaps, not yet quite understood. Mr. Bates found
that the Amazon species used them to thatch the domes
of earth covering the entrances to their subterranean
galleries, the pieces of leaf being carefully covered and
kept in position by a thin layer of grains of earth. In
Nicaragua Mr. Belt found the underground cells full of
a brown flocculent matter, which he considers to be the
gnawed leaves connected by a delicate fungus which
ramifies through the mass and which serves as food for
the larvæ ; and he believes that the leaves are really
gathered as manure-heaps to favour the growth of this
fungus !

When they enter houses, which they often do at
night, the Saübas are very destructive. Once, when
travelling on the Rio Negro, I had bought about a peck
of rice, which was tied up in a large cotton handkerchief
and placed on a bench in a native house where we were
spending the night. The next morning we found about
half the rice on the floor, the remainder having been
carried away by the ants ; and the empty handkerchief
was still on the bench, but with hundreds of neat cuts in
it reducing it to a kind of sieve.[1]

The foraging ants of the genus Eciton are another
remarkable group, especially abundant in the equatorial
forests of America. They are true hunters, and seem
to be continually roaming about the forests in great
bands in search of insect prey. They especially devour
maggots, caterpillars, white ants, cockroaches, and other
soft insects ; and their bands are always accompanied by

[1] For a full and most interesting description of the habits and instincts of
this ant, see Bates' *Naturalist on the Amazons*, 2nd edit. pp. 11-18 ; and
Belt's *Naturalist in Nicaragua*, pp. 71-34.

flocks of insectivorous birds who prey upon the winged insects that are continually trying to escape from the ants. They even attack wasps' nests, which they cut to pieces and then drag out the larvæ. They bite and sting severely, and the traveller who accidentally steps into a horde of them will soon be overrun, and must make his escape as quickly as possible. They do not confine themselves to the ground, but swarm up bushes and low trees, hunting every branch, and clearing them of all insect life. Sometimes a band will enter a house, like the driver ants in Africa, and clear it of cockroaches, spiders, centipedes, and other insects. They seem to have no permanent abode and to be ever wandering about in search of prey, but they make temporary habitations in hollow trees or other suitable places.

Perhaps the most extraordinary of all ants are the blind species of Eciton discovered by Mr. Bates, which construct a covered way or tunnel as they march along. On coming near a rotten log, or any other favourable hunting ground, they pour into all its crevices in search of booty, their covered way serving as. a protection to retire to in case of danger. These creatures, of which two species are known, are absolutely without eyes ; and it seems almost impossible to imagine that the loss of so important a sense-organ can be otherwise than injurious to them. Yet on the theory of natural selection the successive variations by which the eyes were reduced and ultimately lost must all have been useful. It is true they do manage to exist without eyes ; but that is probably because, as sight became more and more imperfect, new instincts or new protective modifications were developed to supply its place, and this does not in any

way account for so wide-spread and invaluable a sense
having become permanently lost, in creatures which still
roam about and hunt for prey very much as do their
fellows who can see.

Special Relations between Ants and Vegetation.—
Attention has recently been called to the very remarkable
relations existing between some trees and shrubs and
the ants which dwell upon them. In the Malay Islands
are several curious shrubs belonging to the Cinchonaceæ,
which grow parasitically on other trees, and whose
swollen stems are veritable ants' nests. When very
young the stems are like small, irregular prickly tubers,
in the hollows of which ants establish themselves; and
these in time grow into irregular masses the size of
large gourds, completely honeycombed with the cells of
ants. In America there are some analagous cases
occurring in several families of plants, one of the most
remarkable being that of certain Melastomas which have
a kind of pouch formed by an enlargement of the petiole
of the leaf, and which is inhabited by a colony of small
ants. The hollow stems of the Cecropias (curious trees
with pale bark and large palmate leaves which are
white beneath) are always tenanted by ants, which make
small entrance holes through the bark; but here there
seems no *special* adaptation to the wants of the insect.
In a species of Acacia observed by Mr. Belt, the thorns
are immensely large and hollow, and are always tenanted
by ants. When young these thorns are soft and full of
a sweetish pulpy substance, so that when the ants first
take possession they find a store of food in their house.
Afterwards they find a special provision of honey-glands
on the leaf-stalks, and also small yellow fruit-like bodies

which are eaten by the ants ; and this supply of food permanently attaches them to the plant. Mr. Belt believes, after much careful observation, that these ants protect the plant they live on from leaf-eating insects, especially from the destructive Saüba ants,—that they are in fact a standing army kept for the protection of the plant! This view is supported by the fact that other plants—Passion-flowers, for example—have honey-secreting glands on the young leaves and on the sepals of the flower-buds which constantly attract a small black ant. If this view is correct, we see that the need of escaping from the destructive attacks of the leaf-cutting ants has led to strange modifications in many plants. Those in which the foliage was especially attractive to these enemies were soon weeded out unless variations occurred which tended to preserve them. Hence the curious phenomenon of insects specially attracted to certain plants to protect them from other insects ; and the existence of the destructive leaf-cutting ant in America will thus explain why these specially modified plants are so much more abundant there than in the Old World, where no ants with equally destructive habits appear to exist.

Wasps and Bees.—These insects are excessively numerous in the tropics, and, from their large size, their brilliant colours, and their great activity, they are sure to attract attention. Handsomest of all, perhaps, are the Scoliadæ, whose large and rather broad hairy bodies, often two inches long, are richly banded with yellow or orange. The Pompilidæ comprise an immense number of large and handsome insects, with rich blue-black bodies and wings and exceedingly long legs. They may often be seen in the forests dragging along large spiders, beetles, or other insects they have captured. Some of the smaller species enter houses and build earthen cells which they store with small green spiders rendered torpid by stinging, to feed the larvæ. The Eumenidæ are beautiful wasps with very long pedunculated bodies, which build papery cones ccvering a few cells in which the eggs are deposited. Among the bees the Xylocopas, or wood-boring bees, are remarkable. They resemble large humble-bees, but have broad, flat, shining bodies, either black or banded with blue; and they often bore large cylindrical holes in the posts of houses. True honey-bees are chiefly remarkable in the East for their large semi-circular combs suspended from the branches of the loftiest trees without any covering. From these exposed nests large quantities of wax and honey are obtained, while the larvæ afford a rich feast to the natives of Borneo, Timor, and other islands where bees abound. They are very pugnacious, and, when disturbed will follow the intruders for miles, stinging severely.

Orthoptera and other Insects.—Next to the butterflies and ants, the insects that are most likely to attract the attention of the stranger in the tropics are the various forms of Mantidæ and Phasmidæ, some of which are remarkable for their strange attitudes and bright colours; while others are among the most singular of known insects, owing to their resemblance to sticks and leaves. The Mantidæ—usually called "praying insects," from their habit of sitting with their long fore-feet held up as if in prayer—are really tigers among insects, lying in wait for their prey, which they seize with their powerful serrated fore-feet. They are usually so coloured as to resemble the foliage among which they live, and as they sit quite motionless, they are not easily perceived.

The Phasmidæ are perfectly inoffensive leaf-eating insects of very varied forms ; some being broad and leaf-like, while others are long and cylindrical so as to resemble sticks, whence they are often called walking-stick insects. The imitative resemblance of some of these insects to the plants on which they live is marvellous.   The true leaf-insects of the East, forming the genus Phyllium, are the size of a moderate leaf, which their large wing-covers and the dilated margins of the head, thorax and legs cause them exactly to resemble.   The veining of the wings, and their green tint, exactly corresponds to that of the leaves of their food-plant ; and as they rest motionless during the day, only feeding at night, they the more easily escape detection.   In Java they are often kept alive on a branch of the guava tree ; and it is a common thing for a stranger, when asked to look at this curious insect, to inquire where it is, and on being told that it is close under his eyes, to maintain that there is no insect at all, but only a branch with green leaves.

The larger wingless stick-insects are often eight inches to a foot long.   They are abundant in the Moluccas ; hanging on the shrubs that line the forest-paths ; and they resemble sticks so exactly, in colour, in the small rugosities of the bark, in the knots and small branches, imitated by the joints of the legs, which are either pressed close to the body, or stuck out at random, that it is absolutely impossible, by the eye alone, to distinguish the real dead twigs which fall down from the trees overhead from the living insects.   The writer has often looked at them in doubt, and has been obliged to use the sense of touch to determine the point. Some are small and slender like the most delicate twigs ; others again have wings ; and it is curious that these wings are often beautifully coloured, generally bright pink, sometimes yellow, and sometimes finely banded with black ; but when at rest these wings.fold up so as to be completely concealed under the narrow wing-covers, and the whole insect is then green or brown, and almost invisible among the twigs or foliage. To increase the resemblance to vegetation, some of these.Phasmas have small green processes in various parts of their bodies looking exactly like moss. These inhabit damp forests both in the Malay islands and in America, and they are so marvellously like moss-grown twigs that the closest examination is needed to satisfy oneself that it is really a living insect we are looking at.

Many of the locusts are equally well disguised, some resembling green leaves, others those that are brown and dead ; and the latter often have small transparent spots on the wings, looking like holes eaten through them. That these disguises deceive their natural enemies is certain, for otherwise the Phasmidæ would soon be exterminated. They are large and sluggish, and very soft and succulent; they have no means of defence or of flight, and they are eagerly devoured by numbers of birds, especially by the numerous cuckoo tribe, whose stomachs are often full of them ; yet numbers of them escape destruction, and this can only be due to their vegetable disguises. Mr. Belt records a curious instance of the actual operation of this kind of defence in a leaf-like locust, which remained perfectly quiescent in the midst of a host of insectivorous ants, which ran over it without finding out that it was an insect and not a leaf ! It might have flown away from them, but it would then instantly have fallen a prey to the numerous birds which always accompany these roaming hordes of ants to feed upon the insects that endeavour to escape. Far more conspicuous than any of these imitative species are the large locusts, with rich crimson or blue-and-black spotted wings. Some of these are nearly a foot in expanse of wings; they fly by day, and their strong spiny legs probably serve as a protection against all the smaller birds. They cannot be said to be common; but when met with they fully satisfy our notions as to the large size and gorgeous colours of tropical insects.

Beetles. — Considering the enormous numbers and endless variety of the beetle tribe that are known to inhabit the tropics, they form by no means so prominent a feature in the animal life of the equatorial zone as we might expect. Almost every entomologist is at first disappointed with them. He finds that they have to be searched for almost as much as at home, while those of large size (except one or two very common species) are rarely met with. The groups which most attract attention from their size and beauty, are the Buprestidæ and the Longicorns. The former are usually smooth insects of an elongate ovate form, with very short legs and antennæ, and adorned with the most glowing metallic tints. They abound on fallen tree-trunks and on foliage, in the hottest sunshine, and are among the most brilliant ornaments of the tropical forests. Some parts of the temperate zone, especially Australia and Chili, abound in Buprestidæ which are equally beautiful; but the largest species are only found within the tropics, those of the Malay islands being the largest of all.

The Longicorns are elegantly shaped beetles, usually with long antennæ and legs, varied in form and structure in an endless variety of ways, and adorned with equally varied colours, spots and markings. Some are large and massive insects three or four inches long, while others are no bigger than our smaller ants. The majority have sober colours, but often delicately marbled, veined, or spotted ; while others are red, or blue, or yellow, or adorned with the richest metallic tints. Their antennæ are sometimes excessively long and graceful, often adorned with tufts of hair, and sometimes pectinated. They especially abound where timber trees have been recently felled in the primeval forests ; and while extensive clearings are in progress their variety seems endless. In such a locality in the island of Borneo, nearly 300 different species were found during one dry season, while the number obtained during eight years' collecting in the whole Malay Archipelago was about a thousand species.

Among the beetles that always attract attention in the tropics are the large, horned, Copridæ and Dynastidæ, corresponding to our dung-beetles. Some of these are of great size, and they are occasionally very abundant. The immense horn-like protuberances on the head and thorax of the males in some of the species are very extraordinary, and, combined with their polished or rugose metallic colours, render them perhaps the most conspicuous of all the beetle tribe. The weevils and their allies are also very interesting, from their immense numbers, endless variety, and the extreme beauty of many of the species. The Anthribidæ, which are especially abundant in the Malay Archipelago, rival the

Longicorns in the immense length of their elegant antennæ; while the diamond beetles of Brazil, the Eupholi of the Papuan islands, and the Pachyrhynchi of the Philippines, are veritable living jewels.

Where a large extent of virgin forest is cut down in the early part of the dry season, and some hot sunny weather follows, the abundance and variety of beetles attracted by the bark and foliage in various stages of drying is amazing. The air is filled with the hum of their wings. Golden and green Buprestidæ are flying about in every direction, and settling on the bark in full sunshine. Green and spotted rose-chafers hum along near the ground ; long-horned Anthribidæ are disturbed at every step ; elegant little Longicorns circle about the drying foliage, while larger species fly slowly from branch to branch. Every fallen trunk is full of life. Strange mottled, and spotted, and rugose Longicorns, endless Curculios, queer-shaped Brenthidæ, velvety brown or steel-blue Cleridæ, brown or yellow or whitish click beetles, (Elaters), and brilliant metallic Carabidæ. Close by, in the adjacent forest, a whole host of new forms are found. Elegant tiger-beetles, leaf-hunting Carabidæ, musk-beetles of many sorts, scarlet Telephori, and countless Chrysomelas Hispas, Coccinellas, with strange Heteromera, and many curious species which haunt fungi, rotten bark or decaying leaves. With such variety and beauty the most ardent entomologist must be fully satisfied ; and when, every now and then, some of the giants of the tropics fall in his way—grand Prionidæ or Lamiidæ several inches long, a massive golden Buprestis, or a monster horned Dynastes—he feels that his most exalted notions of the insect-life of the tropics are at length realized.

Wingless Insects.—Passing on to other orders of insects, the hemiptera, dragon-flies, and true flies hardly call for special remark. Among them are to be found a fair proportion of large and handsome species, but they require much searching after in their special haunts, and seldom attract so much attention as the groups of insects already referred to. More prominent are the wingless tribes, such as spiders, scorpions, and centipedes. The wanderer in the forests often finds the path closed by large webs almost as strong as silk, inhabited by gorgeous spiders with bodies nearly two inches long and legs expanding six inches. Others are remarkable for their hard flat bodies, terminating in horned processes which are sometimes long, slender, and curved like a pair of miniature cow's horns. Hairy terrestrial species of large size are often met with, the largest belonging to the South American genus Mygale, which sometimes actually kill birds, a fact which had been stated by Madame Merian and others, but was discredited till Mr. Bates succeeded in catching one in the act. The small jumping spiders are also noticeable from their immense numbers, variety, and beauty. They frequent foliage and flowers, running about actively in pursuit of small insects; and many of them are so exquisitely coloured as to resemble jewels rather than spiders. Scorpions and centipedes make their presence known to every traveller. In the forests of the Malay islands are huge scorpions of a greenish colour and eight or ten inches long ; while in huts and houses smaller species lurk under boxes and boards, or secrete themselves in almost every article not daily examined. Centipedes of immense size and deadly venom harbour in the thatch of houses and

H

canoes, and will even ensconce themselves under pillows and in beds, rendering a thorough examination necessary before retiring to rest. Yet with moderate precautions there is little danger from these disgusting insects, as may be judged by the fact that during twelve years wanderings in American and Malayan forests the author was never once bitten or stung by them.

General Observations on Tropical Insects.—The characteristics of tropical insects that will most attract the ordinary traveller, are their great numbers, and the large size and brilliant colours often met with. But a more extended observation leads to the conclusion that the average of size is probably no greater in tropical than in temperate zones, and that, to make up for a certain proportion of very large, there is a corresponding increase in the numbers of very small species. The much greater size reached by many tropical insects is no doubt due to the fact, that the supply of food is always in excess of their demands in the larva state, while there is no check from the ever-recurring cold of winter ; and they are thus able to acquire the dimensions that may be on the whole most advantageous to the race, unchecked by the annual or periodical scarcities which in less favoured climates would continually threaten their extinction. The colours of tropical insects are, probably, on the average more brilliant than those of temperate countries, and some of the causes which may have led to this have been discussed in another part of this volume.[1] It is in the tropics that we find most largely developed, whole groups of insects which are unpalatable to almost all insectivorous creatures, and it is among these

[1] Chapters V. and VI.—*The Colours of Animals and Plants.*

that some of the most gorgeous colours prevail. Others obtain protection in a variety of ways; and the amount of cover or concealment always afforded by the luxuriant tropical vegetation is probably a potent agent in permitting a full development of colour.

Birds.—Although the number of brilliantly-coloured birds in almost every part of the tropics is very great, yet they are by no means conspicuous; and as a rule they can hardly be said to add much to the general effect of equatorial scenery. The traveller is almost always disappointed at first with the birds, as he is with the flowers and the beetles; and it is only when, gun in hand, he spends days in the forest, that he finds out how many beautiful living things are concealed by its dense foliage and gloomy thickets. A considerable number of the handsomest tropical birds belong to family groups which are confined to one continent with its adjacent islands; and we shall therefore be obliged to deal for the most part with such large divisions as tribes and orders, by means of which to define the characteristics of tropical bird-life. We find that there are three important orders of birds which, though by no means exclusively tropical, are yet so largely developed there in proportion to their scarcity in extra-tropical regions, that more than any others they serve to give a special character to equatorial ornithology. These are the Parrots, the Pigeons, and the Picariæ, to each of which groups we will devote some attention.

Parrots.—The parrots, forming the order Psittaci of naturalists, are a remarkable group of fruit-eating birds, of such high and peculiar organization that they are often considered to stand at the head of the entire class.

They are pre-eminently characteristic of the intertropical zone, being nowhere absent within its limits (except from absolutely desert regions), and they are generally so abundant and so conspicuous as to occupy among birds the place assigned to butterflies among insects. A few species range far into the temperate zones. One reaches Carolina in North America, another the Magellan Straits in South America; in Africa they only extend a few degrees beyond the southern tropic; in North-Western India they reach 35° North Latitude; but in the Australian region they range farthest towards the pole, being found not only in New Zealand, but as far as the Macquarie Islands in 54° South, where the climate is very cold and boisterous, but sufficiently uniform to supply vegetable food throughout the year. There is hardly any part of the equatorial zone in which the traveller will not soon have his attention called to some members of the parrot tribe. In Brazil, the great blue and yellow or crimson macaws may be seen every evening wending their way homeward in pairs, almost as commonly as rooks with us; while innumerable parrots and parraquets attract attention by their harsh cries when disturbed from some favourite fruit-tree. In the Moluccas and New Guinea, white cockatoos and gorgeous lories in crimson and blue, are the very commonest of birds.

No group of birds—perhaps no other group of animals—exhibits within the same limited number of genera and species, so wide a range and such an endless variety of colour. As a rule parrots may be termed green birds, the majority of the species having this colour as the basis of their plumage relieved by caps, gorgets, bands and wing-spots of other and brighter hues. Yet this general green tint sometimes changes into light or deep blue, as in some macaws; into pure yellow or rich orange, as in some of the American macaw-parrots (*Conurus*) ; into purple, grey, or dove-colour, as in some American, African, and Indian species ; into the purest crimson, as in some of the lories ; into rosy-white and pure white, as in the cockatoos ; and into a deep purple, ashy or black, as in several Papuan, Australian, and Mascarene species. There is in fact hardly a single distinct and definable colour that cannot be fairly matched among the 390 species of known parrots. Their habits, too, are such as to bring them prominently before the eye. They usually feed in flocks ; they are noisy, and so attract attention ; they love gardens, orchards, and open sunny places ; they wander about far in search of food, and towards sunset return homewards in noisy flocks, or in constant pairs. Their forms and motions are often beautiful and attractive. The immensely long tails of the macaws, and the more slender tails of the Indian parraquets; the fine crest of the cockatoos; the swift flight of many of the smaller species, and the graceful motions of the little love-birds and allied forms ; together with their affectionate natures, aptitude for domestication, and powers of mimicry—combine to render them at once the most conspicuous and the most attractive of all the specially tropical forms of bird-life.

The number of species of parrots found in the different divisions of the tropics is very unequal. Africa is by far the poorest; since along with Madagascar and the Mascarene islands, which have many peculiar forms, it scarcely numbers two dozen species. Asia, along with the Malay islands as far as Java and Borneo, is also very poor, with about thirty species. Tropical America is very much richer, possessing about 140 species, among which are many of the largest and most beautiful forms. But of all parts of the globe the tropical islands belonging to the Australian region (from Celebes eastward), together with the tropical parts of Australia, are richest in the parrot tribe, possessing about 150 species, among which are many of the most remarkable and beautiful of the entire group. The whole Australian region, whose extreme limits may be defined by Celebes, the Marquesas, and the New Zealand group, possesses about 200 species of parrots.

Pigeons.—These are such common birds in all temperate countries, that it may surprise many readers to learn that they are nevertheless a characteristic tropical group. That such is the case, however, will be evident from the fact that only sixteen species are known from the whole of the temperate parts of Europe, Asia, and North America, while about 330 species inhabit the tropics. Again, the great majority of the species are found congregated in the equatorial zone, whence they diminish gradually toward the limits of the tropics, and then suddenly fall off in the temperate zones. Yet although they are pre-eminently tropical or even equatorial as a group, they are not, from our present point of view, of much importance, because they are so shy and so generally inconspicuous that in most parts of the tropics an ordinary observer might hardly be aware of their existence. The remark applies especially to America and Africa, where they are neither very abundant nor peculiar ; but in the Eastern hemisphere, and especially in the Malay Archipelago and Pacific islands, they occur in such profusion and present such singular forms and brilliant colours, that they are sure to attract attention. Here we find the extensive group of fruit-pigeons, which, in their general green colours adorned with patches and bands of purple, white, blue, or orange, almost rival the parrot tribe ; while the golden-green Nicobar pigeon, the great crowned pigeons of New Guinea as large as turkeys, and the golden-yellow fruit-dove of the Fijis, can hardly be surpassed for beauty.

Pigeons are especially abundant and varied in tropical archipelagoes ; so that if we take the Malay and Pacific islands, the Madagascar group, and the Antilles or West Indian islands, we find that they possess between them more different kinds of pigeons than all the continental tropics combined. Yet further, that portion of the Malay Archipelago east of Borneo, together with the Pacific islands, is exceptionally rich in pigeons ; and the reason seems to be that monkeys and all other arboreal mammals that devour eggs are entirely absent from this region. Even in South America pigeons are scarce where monkeys are abundant, and *vice versâ ;* so that here we seem to get a glimpse of one of the curious interactions of animals on each other, by which their distribution, their habits, and even their colours may have been influenced ; for the most conspicuous pigeons, whether by colour or by their crests, are all found in countries where they have the fewest enemies.

Picariæ.—The extensive and heterogeneous series of birds now comprised under this term, include most of the fissirostral and scansorial groups of the older naturalists. They may be described as, for the most part, arboreal birds, of a low grade of organization, with weak or abnormally developed feet, and usually less active than the true Passeres or perching birds, of which our warblers, finches, and crows may be taken as the types. The order Picariæ comprises twenty-five families, some of which are very extensive. All are either wholly or mainly tropical, only two of the families—the woodpeckers and the kingfishers—having a few representatives which are permanent residents in the temperate regions; while our summer visitor, the cuckoo, is the sole example in Northern Europe of one of the most abundant and widespread tropical families of birds. Only four of the families have a general distribution over all the warmer countries of the globe—the cuckoos, the kingfishers, the swifts, and the goatsuckers; while two others—the trogons and the woodpeckers—are only wanting in the Australian region, ceasing suddenly at Borneo and Celebes respectively.

Cuckoos.—Whether we consider their wide range, their abundance in genera and species, or the peculiarities of their organization, the cuckoos may be taken as the most typical examples of this extensive order of birds; and there is perhaps no part of the tropics where they do not form a prominent feature in the ornithology of the country. Their chief food consists of soft insects, such as caterpillars, grasshoppers, and the defenceless stick- and leaf-insects; and in search after these they frequent the bushes and lower parts of the forest, and the more open tree-clad plains. They vary greatly in size and appearance, from the small and beautifully metallic golden-cuckoos of Africa, Asia, and Australia, no larger than sparrows, to the pheasant-like ground cuckoo of Borneo, the Scythrops of the Moluccas which almost resembles a hornbill, the Rhamphococcyx of Celebes with its richly-coloured bill, and the Goliath cuckoo of Gilolo with its enormously long and ample tail.

Cuckoos, being invariably weak and defenceless birds, conceal themselves as much as possible among foliage or herbage ; and as a further protection many of them have acquired the coloration of rapacious or combative birds. In several parts of the world cuckoos are coloured exactly like hawks, while some of the small Malayan cuckoos closely resemble the pugnacious drongo-shrikes.

Trogons, Barbets, and Toucans.—Many of the families of Picariæ are confined to the tropical forests, and are remarkable for their varied and beautiful colouring. Such are the trogons of America, Africa, and Malaya, whose dense puffy plumage exhibits the purest tints of rosy-pink, yellow, and white, set off by black heads and a golden-green or rich brown upper surface. Of more slender forms, but hardly less brilliant in colour, are the jacamars and motmots of America, with the bee-eaters and rollers of the East, the latter exhibiting tints of pale blue or verditor-green, which are very unusual. The barbets are rather clumsy fruit-eating birds, found in all the great tropical regions except that of the Austro-Malay islands ; and they exhibit a wonderful variety as well as strange combinations of colours. Those of Asia and Malaya are mostly green, but adorned about the head and neck with patches of

the most vivid reds, blues, and yellows, in endless combinations. The African species are usually black or
greenish-black, with masses of intense crimson, yellow,
or white, mixed in various proportions and patterns;
while the American species combine both styles of
colouring, but the tints are usually more delicate, and
are often more varied and more harmoniously inter-
blended. In the Messrs. Marshall's fine work [1] all the
species are described and figured; and few more instructive examples can be found than are exhibited in
their beautifully-coloured plates, of the endless ways in
which the most glaring and inharmonious colours are
often combined in natural objects with a generally
pleasing result.

We will next group together three families which, although quite distinct, may be said to represent each other
in their respective countries,—the toucans of America, the
plantain-eaters of Africa, and the hornbills of the East
—all being large and remarkable birds which are sure
to attract the traveller's attention. The toucans are the
most beautiful, on account of their large and richly-
coloured bills, their delicate breast-plumage, and the
varied bands of colour with which they are often adorned.
Though feeding chiefly on fruits, they also devour birds'
eggs and young birds; and they are remarkable for the
strange habit of sleeping with the tail laid flat upon
their backs, in what seems a most unnatural and in-
convenient position. What can be the use of their
enormous bills has been a great puzzle to naturalists,
the only tolerably satisfactory solution yet arrived at

[1] *A Monograph of the Capitonidæ or Scansorial Barbets*, by C. F. T.
Marshall and G. F. L. Marshall. 1871.

being that suggested by Mr. Bates,—that it simply enables them to reach fruit at the ends of slender twigs which, owing to their weight and clumsiness, they would otherwise be unable to obtain. At first sight it appears very improbable that so large and remarkable an organ should have been developed for such a purpose ; but we have only to suppose that the original toucans had rather large and thick bills, not unlike those of the barbets (to which group they are undoubtedly allied), and that as they increased in size and required more food, only those could obtain a sufficiency whose unusually large beaks enabled them to reach furthest. So large and broad a bill as they now possess would not be required ; but the development of the bill naturally went on as it had begun, and, so that it was light and handy, the large size was no disadvantage if length was obtained. The plantain-eaters of Africa are less remarkable birds, though adorned with rich colours and elegant crests. The hornbills, though less beautiful than the toucans, are more curious, from the strange forms of their huge bills, which are often adorned with ridges, knobs, or recurved horns. They are bulky and heavy birds, and during flight beat the air with prodigious force, producing a rushing sound very like the puff of a locomotive, and which can sometimes be heard a mile off. They mostly feed on fruits ; and as their very short legs render them even less active than the toucans, the same explanation may be given of the large size of their bills, although it will not account for the curious horns and processes from which they derive their distinctive name. The largest hornbills are more than four feet long, and their laboured noisy flight and

huge bills, as well as their habits of perching on the top
of bare or isolated trees, render them very conspicuous
objects.

The Picariæ comprise many other interesting families;
as, for example, the puff-birds, the todies, and the humming-birds; but as these are all confined to America we
can hardly claim them as characteristic of the tropics
generally. Others, though very abundant in the tropics,
like the kingfishers and the goatsuckers, are too well
known in temperate lands to allow of their being considered as specially characteristic of the equatorial zone.
We will therefore pass on to consider what are the more
general characteristics of the tropical as compared with
the temperate bird-fauna, especially as exemplified
among the true perchers or Passeres, which constitute
about three-fourths of all terrestrial birds.

Passeres.—This great order comprises all our most
familiar birds, such as the thrushes, warblers, tits, shrikes,
flycatchers, starlings, crows, wagtails, larks, and finches.
These families are all more or less abundant in the
tropics; but there are a number of other families which
are almost or quite peculiar to tropical lands and give
a special character to their bird-life. All the peculiarly
tropical families are, however, confined to some definite
portion of the tropics, a number of them being American
only, others Australian, while others again are common
to all the warm countries of the Old World; and it is a
curious fact that there is no single family of this great
order of birds that is confined to the entire tropics, or
that is even especially characteristic of the tropical zone,
like the cuckoos among the Picariæ. The tropical
families of passerine birds being very numerous, and

their peculiarities not easily understood by any but ornithologists, it will be better to consider the series of fifty families of Passeres as one compact group, and endeavour to point out what external peculiarities are most distinctive of those which inhabit tropical countries.

Owing to the prevalence of forests and the abundance of flowers, fruits, and insects, tropical and especially equatorial birds have become largely adapted to these kinds of food ; while the seed-eaters, which abound in temperate lands where grasses cover much of the surface, are proportionately scarce. Many of the peculiarly tropical families are therefore either true insect-eaters or true fruit-eaters, whereas in the temperate zones a mixed diet is more general.

One of the features of tropical birds that will first strike the observer, is the prevalence of crests and of ornamental plumage in various parts of the body, and especially of extremely long or curiously shaped feathers in the tails, tail-coverts, or wings of a variety of species. As examples we may refer to the red paradise-bird, whose middle tail-feathers are like long ribands of whalebone ; to the wire-like tail-feathers of the king bird-of-paradise of New Guinea, and of the wire-tailed manakin of the Amazons ; and to the long waving tail-plumes of the whydah finch of West Africa and paradise-flycatcher of India ; to the varied and elegant crests of the cock-of-the-rock, the king-tyrant, the umbrella-bird, and the six-plumed bird-of-paradise ; and to the wonderful side-plumes of most of the true paradise-birds. In other orders of birds we have such remarkable examples as the racquet-tailed kingfishers of the Moluccas, and the racquet-tailed parrots of

Celebes ; the enormously developed tail-coverts of the peacock and the Mexican trogon ; and the excessive wing-plumes of the argus-pheasant of Malacca and the long-shafted goatsucker of West Africa.

Still more remarkable are the varied styles of coloration in the birds of tropical forests, which rarely or never appear in those of temperate lands. We have intensely lustrous metallic plumage in the jacamars, trogons, humming-birds, sun-birds, and paradise-birds ; as well as in some starlings, pittas or ground thrushes, and drongo-shrikes. Pure green tints occur in parrots, pigeons, green bulbuls, greenlets, and in some tanagers, finches, chatterers, and pittas. These undoubtedly tend to concealment ; but we have also the strange phenomenon of white forest-birds in the tropics, a colour only found elsewhere among the aquatic tribes and in the arctic regions. Thus, we have the bell-bird of South America, the white pigeons and cockatoos of the East, with a few starlings, woodpeckers, kingfishers, and goatsuckers, which are either very light-coloured or in great part pure white.

But besides these strange, and new, and beautiful forms of bird-life, which we have attempted to indicate as characterising the tropical regions, the traveller will soon find that there are hosts of dull and dingy birds, not one whit different, so far as colour is concerned, from the sparrows, warblers, and thrushes of our northern climes. He will however, if observant, soon note that most of these dull colours are protective ; the groups to which they belong frequenting low thickets, or the ground, or the trunks of trees. He will find groups of birds specially adapted to certain modes of tropical life.

Some live on ants upon the ground, others peck minute insects from the bark of trees; one group will devour bees and wasps, others prefer caterpillars; while a host of small birds seek for insects in the corollas of flowers. The air, the earth, the undergrowth, the tree-trunks, the flowers, and the fruits, all support their specially adapted tribes of birds. Each species fills a place in nature, and can only continue to exist so long as that place is open to it; and each has become what it is in every detail of form, size, structure, and even of colour, because it has inherited through countless ancestral forms all those variations which have best adapted it among its fellows to fill that place, and to leave behind it equally well adapted successors.

Reptiles and Amphibia.—Next to the birds, or perhaps to the less observant eye even before them, the abundance and variety of reptiles form the chief characteristic of tropical nature; and the three groups—Lizards, Snakes, and Frogs, comprise all that, from our present point of view, need be noticed.

Lizards.—Lizards are by far the most abundant in individuals and the most conspicuous; and they constitute one of the first attractions to the visitor from colder lands. They literally swarm everywhere. In cities they may be seen running along walls and up palings; sunning themselves on logs of wood, or creeping up to the eaves of cottages. In every garden, road, or dry sandy path, they scamper aside as you walk along. They crawl up trees, keeping at the further side of the trunk and watching the passer-by with the caution of a squirrel. Some will walk up smooth walls with the greatest ease; while in houses the various kinds of Geckos cling to the ceilings, along which they run back downwards in pursuit of flies, holding on by means of their dilated toes with suctorial discs; though sometimes, losing hold, they fall upon the table or on the upturned face of the visitor. In the forests large, flat, and marbled Geckos cling to the smooth trunks; small and active lizards rest on the foliage; while occasionally the larger kinds, three or four feet long, rustle heavily as they move among the fallen leaves.

Their colours vary much, but are usually in harmony with their surroundings and habits. Those that climb about walls and rocks are stone-coloured, and sometimes nearly black; the house lizards are grey or pale-ashy, and are hardly visible on a palm-leaf thatch, or even on a white-washed ceiling. In the forest they are often mottled with ashy-green, like lichen-grown bark. Most of the ground-lizards are yellowish or brown; but some are of beautiful green colours, with very long and slender tails. These are among the most active and lively; and instead of crawling on their bellies like many lizards, they stand well upon their feet and scamper about with the agility and vivacity of kittens. Their tails are very brittle; a slight blow causing them to snap off, when a new one grows, which is, however, not so perfectly formed and completely scaled as the original member. It is not uncommon, when a tail is half broken, for a new one to grow out of the wound, producing the curious phenomenon of a forked tail. There are about 1,300 different kinds of lizards known, the great majority of which inhabit the tropics, and they probably increase in numbers towards the equator. A rich vegetation and a due proportion of moisture and sunshine seem favourable to them, as shown by their great abundance and their varied kinds at Para and in the Aru Islands—places which are nearly the antipodes of each other, but which both enjoy the fine equatorial climate in perfection, and are alike pre-eminent in the variety and beauty of their insect life.

Three peculiar forms of lizard may be mentioned as specially characteristic of the American, African, and Asiatic tropical zones respectively. The iguanas of South America are large arboreal herbivorous lizards of a beautiful green colour, which renders them almost invisible when resting quietly among foliage. They are distinguished by the serrated back, deep dew-lap, and enormously long tail, and are one of the few kinds of lizards whose flesh is considered a delicacy. The chameleons of Africa are also arboreal lizards, and they have the prehensile tail which is more usually found among American animals. They are excessively slow in their motions, and are protected by the wonderful power of changing their colour so as to assimilate it with that of immediately surrounding objects. Like the majority of lizards they are insectivorous, but they are said to be able to live for months without taking food. The dragons or flying lizards of India and the larger Malay islands, are perhaps the most curious and interesting of living reptiles, owing to their power of passing through the air by means of wing-like membranes, which stretch along each side of the body and are expanded by means of slender bony processes from the first six false ribs. These membranes are folded up close to the body when not in use, and are then almost imperceptible ; but when open they form a nearly circular web, the upper surface of which is generally zoned with red or yellow in a highly ornamental manner. By means of this parachute the animal can easily pass from one tree to another for a distance of about thirty feet, descending at first, but as it approaches its destination rising a little so as to reach the tree with its head erect. They are very small, being usually not more than two or three inches long exclusive of the slender tail ; and when the wings are expanded in the sunshine they more resemble some strange insect than one of the reptile tribe.

Snakes.—Snakes are, fortunately, not so abundant or so obtrusive as lizards, or the tropics would be scarcely habitable. At first, indeed, the traveller is disposed to wonder that he does not see more of them, but he will soon find out that there are plenty ; and, if he is possessed by the usual horror or dislike of them, he may think there are too many. In the equatorial zone snakes are less troublesome than in the drier parts of the tropics, although they are probably more numerous and more varied. This is because the country is naturally a vast forest, and the snakes being all adapted to a forest life do not as a rule frequent gardens and come into houses as in India and Australia, where they are accustomed to open and rocky places. One cannot traverse the forest, however, without soon coming upon them. The slender green whip-snakes glide among the bushes, and may often be touched before they are seen. The ease and rapidity with which these snakes pass through bushes, almost without disturbing a leaf, is very curious. More dangerous are the green vipers, which lie coiled motionless upon foliage, where their colour renders it difficult to see them. The writer has often come upon them while creeping through the jungle after birds or insects, and has sometimes only had time to draw back when they were within a few inches of his face. It is startling in walking along a forest path to see a long snake glide away from just where you were going to set down your foot ; but it is perhaps even more alarming to hear a long-drawn heavy slur-r-r, and just to catch a glimpse of a serpent as thick as your leg and an unknown number of feet in length, showing that you must have passed unheeding within a short distance of where it was lying. The smaller pythons are not however dangerous, and they often enter houses to catch and feed upon the rats, and are rather liked by the natives. You will sometimes be told, when sleeping in a native house, that there is a large snake in the roof, and that you need not be disturbed in case you should hear it hunting after its prey. These serpents no doubt sometimes grow to an enormous size, but such monsters are rare. In Borneo, Mr. St. John states that he measured one twenty-six feet long, probably the largest ever measured by a European in the East. The great water-boa of South America is believed to reach the largest size. Mr. Bates measured skins twenty-one feet long, but the largest ever met with by a European appears to be that described by the botanist, Dr. Gardiner, in his *Travels in Brazil*. It had devoured a horse, and was found dead, entangled in the branches of a tree overhanging a river, into which it had been carried by a flood. It was nearly forty feet long. These creatures are said to seize and devour full-sized cattle on the Rio Branco ; and from what is known of their habits this is by no means improbable.

Frogs and Toads.—The only Amphibia that often meet the traveller's eye in equatorial countries are the various kinds of frogs and toads, and especially the elegant tree-frogs. When the rainy season begins, and dried-up pools and ditches become filled with water, there is a strange nightly concert produced by the frogs, some of which croak, others bellow, while many have clanging, or chirruping, and not unmusical notes. In roads and gardens one occasionally meets huge toads six or seven inches long ; but the most abundant and most interesting of the tribe are those adapted for an arboreal life, and hence called tree-frogs. Their toes terminate in discs, by means of which they can cling firmly to leaves and stems. The majority of them are green or brown, and these usually feed at night, sitting quietly during the day so as to be almost invisible, owing to their colour and their moist shining skins so closely resembling vegetable surfaces. Many are beautifully marbled and spotted, and when sitting on leaves resemble large beetles more than frogs, while others are adorned with bright and staring colours ; and these, as Mr. Belt has discovered, have nauseous secretions which render them uneatable, so that they have no need to conceal themselves. Some of these are bright blue, others are adorned with yellow stripes, or have a red body with blue legs. Of the smaller tree-frogs of the tropics there must be hundreds of species still unknown to naturalists.

Mammals—Monkeys.—The highest class of animals, the Mammalia, although sufficiently abundant in all equatorial lands, are those which are least seen by the traveller. There is, in fact, only one group—the monkeys—which are at the same time pre-eminently tropical and which make themselves perceived as one of the aspects of tropical nature. They are to be met with in all the great continents and larger islands, except Australia, New Guinea, and Madagascar, though the latter island possesses the lower allied form of Lemurs ; and they never fail to impress the observer with a sense of the exuberant vitality of the tropics. They are pre-eminently arboreal in their mode of life, and are consequently most abundant and varied where vegetation reaches its maximum development. In the East we find that maximum in Borneo, and in the West African forests ; while in the West the great forest plain of the Amazon stands pre-eminent. It is near the equator only that the great Anthropoid apes, the gorilla, chimpanzee, and orang-utan are found, and they may be met with by any persevering explorer of the jungle. The gibbons, or long-armed apes, have a wider range in the Asiatic continent and in Malaya, and they are more abundant both in species and individuals. Their plaintive howling notes may often be heard in the forests, and they are constantly to be seen sporting at the summits of the loftiest trees, swinging suspended by their long arms, or bounding from tree to tree with incredible agility. They pass through the forest at a height of a hundred feet or more, as rapidly as a deer will travel along the ground beneath them. Other monkeys of various kinds are more abundant and usually less shy ; and in places where fire-arms are not much used they will approach the houses and gambol in the trees undisturbed by the approach of man. The most remarkable of the tailed monkeys of the East is the proboscis monkey of Borneo, whose long fleshy nose gives it an aspect very different from that of most of its allies.

In tropical America monkeys are even more abundant than in the East, and they present many interesting peculiarities. They differ somewhat in dentition and in other structural features from all Old World apes, and a considerable number of them have prehensile tails, a peculiarity never found elsewhere. In the howlers and the spider monkeys the tail is very long and powerful, and by twisting the extremity round a branch the animal can hang suspended as easily as other monkeys can by their hands. It is, in fact, a fifth hand, and is constantly used to pick up small objects from the ground. The most remarkable of the American monkeys are the howlers, whose tremendous roaring exceeds that of the lion or the bull, and is to be heard frequently at morning and evening in the primeval forests. The sound is produced by means of a large, thin, bony vessel in the throat, into which air is forced ; and it is very remarkable that this one group of monkeys should possess an organ not found in any other monkey or even in any other mammal, apparently for no other purpose than to be able to make a louder noise than the rest. The only other monkeys worthy of special attention are the marmosets, beautiful little creatures with crests, whiskers, or manes ; in outward form resembling squirrels, but with a very small monkey-like face. They are either black, brown, reddish, or nearly white in colour, and are the smallest of the monkey tribe, some of them being only about six inches long exclusive of the tail.

Bats.—Almost the only other order of mammals that

is specially and largely developed in the tropical zone is that of the Chiroptera or bats ; which becomes suddenly much less plentiful when we pass into the temperate regions, and still more rare towards the colder parts of it, although a few species appear to reach the Arctic circle. The characteristics of the tropical bats are their great numbers and variety, their large size, and their peculiar forms or habits. In the East those which most attract the traveller's attention are the great fruit-bats, or flying-foxes as they are sometimes called, from the rusty colour of the coarse fur and the fox-like shape of the head. These creatures may sometimes be seen in immense flocks which take hours to pass by, and they often devastate the fruit plantations of the natives. They are often five feet across the expanded wings, with the body of a proportionate size ; and when resting in the daytime on dead trees, hanging head downwards, the branches look as if covered with some monster fruits. The descendants of the Portuguese in the East use them for food, but all the native inhabitants reject them.

In South America there is a group of bats which are sure to attract attention. These are the vampyres, several of which are blood-sucking species, which abound in most parts of tropical America and are especially plentiful in the Amazon Valley. Their carnivorous propensities were once discredited, but are too well authenticated. Horses and cattle are often bitten, and are found in the morning covered with blood ; and repeated attacks weaken and ultimately destroy them. Some persons are especially subject to the attacks of these bats ; and as native huts are never sufficiently close to keep them out, these unfortunate individuals are obliged to sleep completely muffled up, in order to avoid being made seriously ill or even losing their lives. The exact manner in which the attack is made is not positively known, as the sufferer never feels the wound. The present writer was once bitten on the toe, which was found bleeding in the morning from a small round hole from which the flow of blood was not easily stopped. On another occasion, when his feet were carefully covered up, he was bitten on the tip of the nose, only awaking to find his face streaming with blood. The motion of the wings fans the sleeper into a deeper slumber, and renders him insensible to the gentle abrasion of the skin either by teeth or tongue. This ultimately forms a minute hole, the blood flowing from which is sucked or lapped up by the hovering vampyre. The largest South American bats, having wings from two to two-and-half feet in expanse, are fruit-eaters like the Pteropi of the East, the true blood-suckers being small or of medium size and varying in colour in different localities. They belong to the genus *Phyllostoma*, and have a tongue with horny papillæ at the end ; and it is probably by means of this that they abrade the skin and produce a small round wound. This is the account given by Buffon and Azara, and there seems now little doubt that it is correct.

Beyond these two great types—the monkeys and the bats—we look in vain among the varied forms of mammalian life for any that can be said to be distinctive of the tropics as compared with the temperate regions. Many peculiar groups are tropical, but they are in almost every case confined to limited portions of the tropical zones, or are rare in species or individuals. Such are the lemurs in Africa, Madagascar, and Southern Asia ; the tapirs of America and Malaya; the rhinoceroses and elephants of Africa and Asia; the cavies and the sloths of America; the scaly ant-eaters of Africa and Asia; but none of these are sufficiently numerous to come often before the traveller so as to affect his general ideas of the aspects of tropical life, and they are, therefore, out of place in such a sketch of those aspects as we are here attempting to lay before our readers.

Summary of the Aspects of Animal Life in the Tropics.—We will now briefly summarize the general aspects of animal life as forming an ingredient in the scenery and natural phenomena of the equatorial regions. Most prominent are the butterflies, owing to their numbers, their size, and their brilliant colours; as well as their peculiarities of form, and the slow and majestic flight of many of them. In other insects, the large size, and frequency of protective colours and markings are prominent features; together with the inexhaustible profusion of the ants and other small insects. Among birds the parrots stand forth as the pre-eminent tropical group, as do the apes and monkeys among mammals; the two groups having striking analogies, in the prehensile hand and the power of imitation. Of reptiles, the two most prominent groups are the lizards and the frogs; the snakes, though equally abundant, being much less obtrusive.

Animal life is, on the whole, far more abundant and more varied within the tropics than in any other part of the globe, and a great number of peculiar groups are found there which never extend into temperate regions. Endless eccentricities of form, and extreme richness of colour are its most prominent features ; and these are manifested in the highest degree in those equatorial lands where the vegetation acquires its greatest beauty and its fullest development. The causes of these essentially tropical features are not to be found in the comparatively simple influence of solar light and heat, but rather in the uniformity and permanence with which these and all other terrestrial conditions have acted ; neither varying prejudicially throughout the year, nor having undergone any important change for countless past ages. While successive glacial periods have devastated the temperate zones, and destroyed most of the larger and more specialized forms which during more favourable epochs had been developed, the equatorial lands must always have remained thronged with life ; and have been unintermittingly subject to those complex influences of organism upon organism, which seem the main agents in developing the greatest variety of forms and filling up every vacant place in nature. A constant struggle against the vicissitudes and recurring severities of climate must always have restricted the range of effective animal variation in the temperate and frigid zones, and have checked all such developments of form and colour as were in the least degree injurious in themselves, or which co-existed with any constitutional incapacity to resist great changes of temperature or other unfavourable conditions. Such disadvantages were not experienced in the equatorial zone. The struggle for existence as against the forces of nature was there always less severe,—food was there more abundant and more regularly supplied,—shelter and concealment were at all times more easily obtained ; and almost the only physical changes experienced, being dependent on cosmical or geological changes, were so slow, that variation and natural selection were always able to keep the teeming mass of organisms in nicely balanced harmony with the changing physical conditions. The equatorial zone, in short, exhibits to us the result of a comparatively continuous and unchecked development of organic forms ; while in the temperate regions, there have been a series of periodical checks and extinctions of a more or less disastrous nature, necessitating the commencement of the work of development in certain lines over and over again. In the one, evolution has had a fair chance ; in the other it has had countless difficulties thrown in its way. The equatorial regions are then, as regards their past and present life history, a more ancient world than that represented by the temperate zones, a world in which the laws which have governed the progressive development of life have operated with comparatively little check for countless ages, and have resulted in those infinitely varied and beautiful forms—those wonderful eccentricities of structure, of function, and of instinct— that rich variety of colour, and that nicely balanced harmony of relations—which delight and astonish us in the animal productions of all tropical countries.

IV.

HUMMING-BIRDS:

AS ILLUSTRATING THE LUXURIANCE OF TROPICAL NATURE.

Structure—Colours and Ornaments—Display of Ornaments by the Male—
Descriptive Names—The Motions and Habits of Humming-birds—Food
—Nests—Geographical Distribution and Variation—Humming-birds of
Juan Fernandez as illustrating Variation and Natural Selection—The
relations and affinities of Humming-birds—How to determine doubtful
affinities—Resemblances of Swifts and Humming-birds—Differences
between Sun-birds and Humming-birds—Conclusion.

THERE are now about ten thousand different kinds of birds known to naturalists, and these are classed in one hundred and thirty families which vary greatly in extent, some containing a single species only, while others comprise many hundreds. The two largest families are those of the warblers, with more than six hundred, and the finches with more than five hundred species, spread over the whole globe ; the hawks and the pigeons, also spread over the whole globe, number about three hundred and thirty, and three hundred and sixty species respectively ; while the diminutive humming-birds, confined to one hemisphere, consist of about four hundred different species. They are thus, as regards the number of distinct kinds collected in a limited area, the most remarkable of all the families of birds. It may, however, very reasonably be asked, whether the four hundred species of humming-birds above alluded to are really all distinct—as distinct on the average as the ten thousand species of birds are from each other. We reply that they certainly are perfectly distinct species which never intermingle ; and their differences do not consist in colour only, but in peculiarities of form, of structure, and of habits ; so that they have to be classed in more than a hundred distinct genera or systematic groups of species, these genera being really as unlike each other as stonechats and nightingales, or as partridges and blackcocks. The figures we have quoted, as showing the proportion of birds in general to humming-birds, thus represent real facts ; and they teach us that these small and in some respects insignificant birds, constitute an important item in the animal life of the globe.

Humming-birds are, in many respects, unusually interesting and instructive. They are highly peculiar in form, in structure, and in habits, and are quite unrivalled as regards variety and beauty. Though the name is familiar to every one, few but naturalists are acquainted with the many curious facts in their history, or know how much material they afford for admiration and study. It is proposed, therefore, to give a brief and popular account of the form, structure, habits, distribution, and affinities, of this remarkable family of birds, as illustrative of the teeming luxuriance of tropical nature, and as throwing light on some of the most interesting problems of natural history.

Structure.—The humming-birds form one compact

family named Trochilidæ. They are all small birds, the largest known being about the size of a swallow, while the smallest are minute creatures whose bodies are hardly larger than a humble-bee. Their distinguishing features are excessively short legs and feet, very long and pointed wings, a long and slender bill, and a long extensible tubular tongue ; and these characters are found combined in no other birds. The feet are exceedingly small and delicate, often beautifully tufted with down, and so short as to be hardly visible beyond the plumage. The toes are placed as in most birds, three in front and one behind, and have very strong and sharply curved claws ; and the feet serve probably to cling to a perch rather than to give any movement to the body. The wings are long and narrow, but strongly formed ; and the first quill is the longest, a peculiarity found in hardly any other birds but a few of the swifts. The bill varies greatly in length, but is always long, slender, and pointed, the upper mandible being the widest and lapping over the lower at each side, thus affording complete protection to the delicate tongue the perfect action of which is essential to the bird's existence. The humming-bird's tongue is very long, and is capable of being greatly extended beyond the beak and rapidly drawn back, by means of muscles which are attached to the hyoid or tongue-bones, and bend round over the back and top of the head to the very forehead, just as in the wood-peckers. The two blades or laminæ, of which the tongues of birds usually seem to be formed, are here greatly lengthened, broadened out, and each rolled up ; so as to form a complete double tube connected down the middle, and with the outer edges in contact but not united. The extremities of the tubes are, however, flat and fibrous. This tubular and retractile tongue enables the bird to suck up honey from the nectaries of flowers, and also to capture small insects ; but whether the latter pass down the tubes, or are entangled in the fibrous tips and thus draw back into the gullet, is not known. The only other birds with a similar tubular tongue are the sun-birds of the East, which however, as we shall presently explain, have no affinity whatever with the humming-birds.

Colours and Ornaments.—The colours of these small birds are exceedingly varied and exquisitely beautiful. The basis of the colouring may be said to be green, as in parrots ; but whereas in the latter it is a silky green, in humming-birds it is always metallic. The majority of the species have some green about them, especially on the back ; but in a considerable number rich blues, purples, and various shades of red are the prevailing tints. The greater part of the plumage has more or less of a metallic gloss, but there is almost always some part which has an intense lustre, as if actually formed of scales of burnished metal. A gorget, covering the greater part of the neck and breast, most commonly displays this vivid colour ; but it also frequently occurs on the head, on the back, on the tail-coverts above or below, on the upper surface of the tail, on the shoulders or even the quills. The hue of every precious stone and the lustre of every metal is here represented ; and such terms as topaz, amethyst, beryl, emerald, garnet, ruby, sapphire ; golden, golden-green, coppery, fiery, glowing, iridescent, refulgent, celestial, glittering, shining, are constantly used to name or describe the different species.

No less remarkable than the colours are the varied developments of plumage with which these birds are adorned. The head is often crested in a variety of ways ; either a simple flat crest, or with radiating feathers, or diverging into two horns, or spreading laterally like wings, or erect and bushy, or recurved and pointed like that of a plover. The throat and breast are usually adorned with broad scale-like feathers, or these diverge into a tippet, or send out pointed collars, or elegant frills of long and narrow plumes tipped with metallic spots of various colours. But the tail is even a more varied and beautiful ornament, either short and rounded, but pure white or some other strongly contrasted tint ; or with short pointed feathers forming a star ; or with the three outer feathers on each side long and tapering to a point ; or larger, and either square, or round, or deeply forked, or acutely pointed ; or with the two middle feathers excessively long and narrow ; or with the tail very long and deeply forked, with broad and richly-coloured feathers ; or with the two outer feathers wire-like and having broad spoon-shaped tips. All these ornaments, whether of the head, neck, breast or tail, are invariably coloured in some effective or brilliant manner, and often contrast strikingly with the rest of the plumage. Again, these colours often vary in tint according to the direction in which they are seen. In some species they must be looked at from above, in others from below ; in some from the front, in others from behind, in order to catch the full glow of the metallic lustre ; hence, when the birds are seen in their native haunts, the colours come and go and change with their motions, so as to produce a startling and beautiful effect.

The bill differs greatly in length and shape, being either straight or gently curved, in some species bent like a sickle, in others turned up like the bill of the avoset. It is usually long and slender, but in one group is so enormously developed that it is nearly the same length as the rest of the bird. The legs, usually little seen, are in some groups adorned with globular tufts of white, brown, or black down, a peculiarity possessed by no other birds. The reader will now be in a position to understand how the four hundred species of humming-birds may be easily distinguished, by the varied combinations of the characters here briefly enumerated, together with many others of less importance. One group of birds will have a short round tail, with crest and long neck-frill; another group a deeply-forked broad tail, combined with glowing crown and gorget; one is both bearded and crested; others have a luminous back and pendent neck-plumes; and in each of these groups the species will vary in combinations of colour, in size, and in the proportions of the ornamental plumes, so as to produce an unmistakable distinctness; while, without any new developments of form or structure, there is room for the discovery of hundreds more of distinct kinds of humming-birds.

Descriptive Names.—The name we usually give to the birds of this family is derived from the sound of their rapidly-moving wings, a sound which is produced by the largest as well as by the smallest member of the group. The Creoles of Guiana similarly call them Bourdons or hummers. The French term, Oiseau-mouche, refers to their small size; while Colibri is a native name which has come down from the Carib inhabitants of the West

K

Indies. The Spaniards and Portuguese call them by
more poetical names, such as Flower-peckers, Flower-
kissers, Myrtle-suckers—while the Mexican and Peruvian
names show a still higher appreciation of their beauties,
their meaning being rays of the sun, tresses of the day-
star, and other such appellations. Even our modern
naturalists, while studying the structure and noting the
peculiarities of these living gems, have been so struck
by their inimitable beauties that they have endeavoured to
invent appropriate English names for the more beautiful
and remarkable genera. Hence we find in common use
such terms as Sun-gems, Sun-stars, Hill-stars, Wood-stars,
Sun-angels, Star-throats, Comets, Coquettes, Flame-
bearers, Sylphs, and Fairies ; together with many others
derived from the character of the tail or the crests.

The Motions and Habits of Humming-birds.—Let us
now consider briefly, the peculiarities of flight, the motions,
the food, the nests, and general habits of the humming-
birds, quoting the descriptions of those modern naturalists
who have personally observed them. Their appearance,
remarks Professor Alfred Newton, is entirely unlike that
of any other bird :—" One is admiring some brilliant and
beautiful flower, when between the blossom and one's
eye suddenly appears a small dark object, suspended as
it were between four short black threads meeting each
other in a cross. For an instant it shows in front of the
flower ; again another instant, and emitting a momentary
flash of emerald and sapphire light, it is vanishing,
lessening in the distance, as it shoots away, to a speck
that the eye cannot take note of." Audubon observes
that the Ruby Humming-birds pass through the air in
long undulations, but the smallness of their size precludes

the possibility of following them with the eye further than fifty or sixty yards, without great difficulty. A person standing in a garden by the side of a common althæa in bloom, will hear the humming of their wings and see the little birds themselves within a few feet of him one moment, while the next they will be out of sight and hearing. Mr. Gould, who visited North America in order to see living humming-birds while preparing his great work on the family, remarks, that the action of the wings reminded him of a piece of machinery acted upon by a powerful spring. When poised before a flower, the motion is so rapid that a hazy semicircle of indistinctness on each side of the bird is all that is perceptible. Although many short intermissions of rest are taken, the bird may be said to live in the air—an element in which it performs every kind of evolution with the utmost ease, frequently rising perpendicularly, flying backward, pirouetting or dancing off, as it were, from place to place, or from one part of a tree to another, sometimes descending, at others ascending. It often mounts up above the towering trees, and then shoots off like a little meteor at a right angle. At other times it gently buzzes away among the little flowers near the ground ; at one moment it is poised over a diminutive weed, at the next it is seen at a distance of forty yards, whither it has vanished with the quickness of thought.

The Rufous Flame-bearer, an exquisite species found on the west coast of North America, is thus described by Mr. Nuttall :—" When engaged in collecting its accustomed sweets, in all the energy of life, it seemed like a breathing gem, a magic carbuncle of flaming fire,

stretching out its glorious ruff as if to emulate the sun
itself in splendour." The Sappho Comet, whose long
forked tail barred with crimson and black renders it one
of the most imposing of humming-birds, is abundant
in many parts of the Andes; and Mr. Bonelli tells us
that the difficulty of shooting them is very great from
the extraordinary turns and evolutions they make when
on the wing; at one instant darting headlong into a
flower, at the next describing a circle in the air with
such rapidity that the eye, unable to follow the movement, loses sight of the bird until it again returns to
the flower which at first attracted its attention. Of the
little Vervain humming-bird of Jamaica, Mr. Gosse
writes:—"I have sometimes watched with much delight
the evolutions of this little species at the Moringa-
tree.[1] When only one is present, he pursues the round of
the blossoms soberly enough. But if two are at the tree,
one will fly off, and suspend himself in the air a few
yards distant; the other presently starts off to him, and
then, without touching each other, they mount upwards
with strong rushing wings, perhaps for five hundred
feet. They then separate, and each starts diagonally
towards the ground like a ball from a rifle, and wheeling
round comes up to the blossoms again as if it had not
moved away at all. The figure of the smaller humming-
birds on the wing, their rapidity, their wavering course,
and their whole manner of flight are entirely those of an
insect." Mr. Bates remarks, that on the Amazons
during the cooler hours of the morning and from four

[1] Sometimes called the horse-radish tree. It is the *Moringa pterygosperma*,
a native of the East Indies, but commonly cultivated in Jamaica. It has
yellow flowers.

to six in the afternoon humming-birds are to be seen
whirring about the trees by scores ; their motions being
unlike those of any other birds. They dart to and fro
so swiftly that the eye can scarcely follow them, and
when they stop before a flower it is only for a few
moments. They poise themselves in an unsteady manner,
their wings moving with inconceivable rapidity, probe
the flower, and then shoot off to another part of the
tree. They do not proceed in that methodical manner
which bees follow, taking the flowers seriatim, but skip
about from one part of the tree to another in the most
capricious way. Mr. Belt remarks on the excessive
rapidity of the flight of the humming-bird giving it a
sense of security from danger, so that it will approach
a person nearer than any other bird, often hovering
within two or three yards (or even one or two feet) of
one's face. He watched them bathing in a small pool
in the forest, hovering over the water, turning from side
to side by quick jerks of the tail ; now showing a throat
of gleaming emerald, now shoulders of glistening
amethyst ; then darting beneath the water, and rising instantly, throw off a shower of spray from their quivering
wings, and again fly up to an overhanging bough and
commence to preen their feathers. All humming-birds
bathe on the wing, and generally take three or four dips,
hovering between times about three or four inches above
the surface. Mr. Belt also remarks on the immense
numbers of humming-birds in the forests, and the great
difficulty of seeing them ; and his conclusion is, that in
the part of Nicaragua where he was living they equalled
in number all the rest of the birds together, if they did
not greatly exceed them.

The extreme pugnacity of humming-birds has been noticed by all observers. Mr. Gosse describes two meeting and chasing each other through the labyrinths of twigs and flowers till, an opportunity occurring, the one would dart with seeming fury upon the other, and then, with a loud rustling of their wings, they would twirl together, round and round, till they nearly came to the earth. Then they parted, and after a time another tussle took place. Two of the same species can hardly meet without an encounter, while in many cases distinct species attack each other with equal fury. Mr. Salvin describes the splendid Eugenes fulgens attacking two other species with as much ferocity as its own fellows. One will knock another off its perch, and the two will go fighting and screaming away at a pace hardly to be followed by the eye. Audubon says they attack any other birds that approach them, and think nothing of assaulting tyrant-shrikes and even birds of prey that come too near their home.

Display of Ornaments by the Male.—It is a well-known fact, that when male birds possess any unusual ornaments, they take such positions or perform such evolutions as to exhibit them to the best advantage while endeavouring to attract or charm the females or in rivalry with other males. It is therefore probable that the wonderfully varied decorations of humming-birds, whether burnished breast-shields, resplendent tail, crested head, or glittering back, are thus exhibited ; but almost the only actual observation of this kind is that of Mr. Belt, who describes how two males of the Florisuga mellivora displayed their ornaments before a female bird. One would shoot up like a rocket, then, suddenly expanding the snow-white tail like an inverted parachute, slowly descend in front of her, turning round gradually to show off both back and front. The expanded white tail covered more space than all the rest of the bird, and was evidently the grand feature of the performance. Whilst one was descending the other would shoot up and come slowly down expanded.[1]

Food.—The food of humming-birds has been a matter of much controversy. All the early writers down to Buffon believed that they lived solely on the nectar of flowers ; but since that time every close observer of their habits maintains that they feed largely, and in some cases wholly, on insects. Azara observed them on the La Plata in winter taking insects out of the webs of spiders at a time and place where there were no flowers. Bullock, in Mexico, declares that he saw them catch small butterflies, and that he found many kinds of insects in their stomachs. Waterton made a similar statement. Hundreds and perhaps thousands of specimens have since been dissected by collecting naturalists, and in almost every instance their stomachs have been found full of insects, sometimes, but not generally, mixed with a proportion of honey. Many of them in fact may be seen catching gnats and other small insects just like fly-catchers, sitting on a dead twig over water, darting off for a time in the air, and then returning to the twig. Others come out just at dusk, and remain on the wing, now stationary, now darting about with the greatest rapidity, imitating in a limited space the evolutions of the goatsuckers, and evidently for the same end and purpose. Mr. Gosse also remarks :—" All the hum-

---

[1] *The Naturalist in Nicaragua*, p. 112.

ming-birds have more or less the habit, when in flight, of pausing in the air and throwing the body and tail into rapid and odd contortions.   This is most observable in the Polytmus, from the effect that such motions have on the long feathers of the tail.   That the object of these quick turns is the capture of insects, I am sure, having watched one thus engaged pretty close to me.   I observed it carefully, and distinctly saw the minute flies in the air which it pursued and caught, and heard repeatedly the snapping of the beak.   My presence scarcely disturbed it, if at all."

There is also an extensive group of small brown humming-birds, forming the sub-family Phaëthornithinæ, which rarely or never visit flowers, but frequent the shady recesses of the forest, where they hunt for minute insects.   They dart about among the foliage, and visit in rapid succession every leaf upon a branch, balancing themselves vertically in the air, passing their beaks closely over the under-surface of each leaf, and thus capturing, no doubt, any small insects that may lurk there.   While doing this, the two long feathers of the tail have a vibrating motion, serving apparently as a rudder, to assist them in performing the delicate operation.   Others search up and down stems and dead sticks in the same manner, every now and then picking off something, exactly as a bush-shrike or a tree-creeper does, with the difference that the humming-bird is constantly on the wing ; while the remarkable Sickle-bill is said to probe the scale-covered stems of palms and tree-ferns to obtain its insect food.

It is a well-known fact that although humming-birds are easily tamed, they cannot be preserved long in captivity, even in their own country, when fed only on syrup. Audubon states, that when thus fed they only live a month or two and die apparently starved ; while if kept in a room whose open windows are covered with a fine net, so as to allow small insects to enter, they have been kept for a whole year without any ill-effects. Another writer, Mr. Webber, captured and tamed a number of the Ruby-throat in the United States. He found that when fed for three weeks on syrup they drooped, but after being let free for a day or two they would return to the open cage for more of the syrup. Some which had been thus tamed and set free, returned the following year, and at once flew straight to the remembered little cup of sweets. Mr. Gosse in Jamaica also kept some in captivity, and found the necessity of giving them insect food ; and he remarks that they were very fond of a small ant that swarmed on the syrup with which they were fed. It is strange that, with all this previous experience and information, those who have attempted to bring live humming-birds to this country have fed them exclusively on syrup ; and the weakness produced by this insufficient food has no doubt been the chief cause of their death on, or very soon after, arrival. A box of ants would not be difficult to bring as food for them ; but even finely-chopped meat or yolk of egg would probably serve, in the absence of insects, to supply the necessary proportion of animal food.

Nests.—The nests of the humming-birds are, as might be expected, beautiful objects, some being no larger inside than the half of a walnut-shell. These small cup-shaped nests are often placed in the fork of a branch, and the outside is sometimes beautifully decorated with pieces of lichen, the body of the nest being formed of cottony substances and the inside lined with the finest and most silky fibres. Others suspend their nests to creepers hanging over water, or even over the sea ; and the Pichincha humming-bird once attached its nest to a straw-rope hanging from the roof of a shed. Others again build nests of a hammock-form attached to the face of rocks by spiders' web ; while the little forest-haunting species fasten their nests to the points or to the under-sides of palm-leaves or other suitable foliage. They lay only one or two white eggs.

Geographical Distribution and Variation. — Most persons know that humming-birds are found only in America ; but it is not so generally known that they are almost exclusively tropical birds, and that the few species that are found in the temperate (northern and southern) parts of the continent are migrants, which retire in the winter to the warmer lands near or within the tropics. In the extreme north of America two species are regular summer visitants, one on the east and the other on the west of the Rocky Mountains. On the east the common N. American or Ruby-throated humming-bird extends through the United States and Canada, and as far as 57° north latitude, or considerably north of Lake Winnipeg ; while the milder climate of the west coast allows the Rufous Flame-bearer to extend its range to beyond Sitka to the parallel of 61°. Here they spend the whole summer, and breed, being found on the Columbia River in the latter end of April, but retire to Mexico in the winter. Supposing that those which go furthest north do not return further south than the borders of the tropics, these little birds must make a journey of full three thousand miles each spring and autumn. The antarctic humming-bird visits the inhospitable shores of Tierra-del-Fuego, where it has been seen visiting the flowers of fuchsias in a snow-storm, while it spends the winter in the warmer parts of Chili and Bolivia.

In the south of California and in the Central United States three or four other species are found in summer ; but it is only when we enter the tropics that the number of different kinds becomes considerable. In Mexico there are more than thirty species, while in the southern parts of Central America there are more than double that number. As we go on towards the equator they become still more numerous, till they reach their maximum in the equatorial Andes. They especially abound in the mountainous regions ; while the luxuriant forest plains of the Amazons, in which so many other forms of life reach their maximum, are very poor in humming-birds. Brazil, being more hilly and with more variety of vege- tation, is richer, but does not equal the Andean valleys, plateaux, and volcanic peaks. Each separate district of the Andes has its peculiar species and often its peculiar genera, and many of the great volcanic mountains possess kinds which are confined to them. Thus, on the great mountain of Pichincha there is a peculiar species found at an elevation of about fourteen thousand feet only ; while an allied species on Chimborazo ranges from fourteen thousand feet to the limits of perpetual snow at sixteen thousand feet elevation. It frequents a beautiful yellow-flowered alpine shrub belonging to the Asteraceæ. On the extinct volcano of Chiriqui in Veragua a minute humming-bird, called the little Flame- bearer, has been only found inside the crater. Its scaled gorget is of such a flaming crimson that, as Mr. Gould remarks, it seems to have caught the last spark from the volcano before it was extinguished.

Not only are humming-birds found over the whole extent of America, from Sitka to Tierra-del-Fuego, and from the level of the sea to the snow-line on the Andes, but they inhabit many of the islands at a great distance from the mainland. The West Indian islands possess fifteen distinct species belonging to eight different genera, and these are so unlike any found on the continent that five of these genera are peculiar to the Antilles. Even the Bahamas, so close to Florida, possess two peculiar species. The small group of islands called Tres Marias, about sixty miles from the west coast of Mexico, has a peculiar species. More remarkable are the two humming-birds of Juan Fernandez, situated in the Pacific Ocean, four hundred miles west of Valparaiso in Chili, one of these being peculiar ; while another species inhabits the little island Mas-afuera, ninety miles further west. The Galapagos, though very little further from the mainland and much more extensive, have no humming-birds; neither have the Falkland islands, and the reason seems to be that both these groups are deficient in forest, and in fact have hardly any trees or large shrubs, while there is a great paucity of flowers and of insect life.

Humming-birds of Juan Fernandez as illustrating Variation and Natural Selection.—The three species which inhabit Juan Fernandez and Mas-afuera present certain peculiarities of great interest. They form a distinct genus, Eustephanus, one species of which inhabits Chili as well as the island of Juan Fernandez. This, which may be termed the Chilian species, is greenish in

both sexes, whereas in the two species peculiar to the
islands the males are red or reddish-brown, and the
females green. The two red males differ very slightly
from each other, but the three green females differ considerably ; and the curious point is, that the female in
the smaller and more distant island somewhat resembles
the same sex in Chili, while the female of the Juan
Fernandez species is very distinct, although the males of
the two islands are so much alike. As this forms a
comparatively simple case of the action of the laws of
variation and natural selection, it will be instructive to
see if we can picture to ourselves the process by which
the changes have been brought about. We must first
go back to an unknown but rather remote period, just
before any humming-birds had reached these islands.
At that time a species of this peculiar genus, Eustephanus,
must have inhabited Chili ; but we must not be sure
that it was identically the same as that which is now
found there, because we know that species are always
undergoing change to a greater or less degree. After
perhaps many failures, one or more pairs of the Chilian
bird got blown across to Juan Fernandez, and finding
the country favourable, with plenty of forests and a fair
abundance of flowers and insects, they rapidly increased
and permanently established themselves on the island.
They soon began to change colour, however, the male
getting a tinge of reddish-brown, which gradually
deepened into the fine colour now exhibited by the two
insular species, while the female, more slowly, changed
to white on the under-surface and on the tail, while the
breast-spots became more brilliant. When the change
of colour was completed in the male, but only partially

so in the female, a further emigration westward took place to the small island Mas-afuera, where they also established themselves. Here, however, the change begun in the larger island appears to have been checked, for the female remains to this day intermediate between the Juan Fernandez and the Chilian forms. More recently, the parent form has again migrated from Chili to Juan Fernandez, where it still lives side by side with its greatly changed descendant.[1] Let us now see how far these facts are in accordance with the general laws of variation, and with those other laws which I have endeavoured to show regulate the development of colour.[2]

The amount of variation which is likely to occur in a species will be greatly influenced by two factors—the occurrence of a change in the physical conditions, and the average abundance or scarcity of the individuals composing the species. When from these or other causes variation occurs, it may become fixed as a variety or a race, or may go on increasing to a certain extent, either from a tendency to vary along certain special lines induced by local or physiological causes, or by the continued survival and propagation of all such varieties as are beneficial to the race. After a certain time a balance will be arrived at, either by the limits of useful variation in this one direction having been reached, or by the species becoming harmoniously adapted to all the surrounding conditions ; and without some change in these

---

[1] In the preceding account of the probable course of events in peopling these islands with humming-birds, I follow Mr. Sclater's paper on the *Land Birds of Juan Fernandez,—Ibis*, 1871, p. 183. In what follows, I give my own explanation of the probable causes of the change.

[2] See *Macmillan's Magazine*, Sept. 1867, " On the Colours of Animals and Plants," and Chapters V. and VI. of the present volume.

conditions the specific form may then remain unaltered for a very long time ; whence arises the common impression of the fixity of species. Now in a country like Chili, forming part of a great continent very well stocked with all forms of organic life, the majority of the species would be in a state of stable equilibrium ; the most favourable variations would have been long ago selected ; and the numbers of individuals in each species would be tolerably constant, being limited by the numerous other forms whose food and habits were similar, or which in any way impinged upon its sphere of existence. We may, therefore, assume that the Chilian humming-bird which migrated to Juan Fernandez was a stable form, hardly if at all different from the existing species which is termed Eustephanus galeritus. On the island it met with very changed but highly favourable conditions,—an abundant shrubby vegetation and a tolerably rich flora ; less extremes of climate than on the mainland ; and, most important of all, absolute freedom from the competition of rival species. The flowers and their insect inhabitants were all its own ; there were no snakes or mammalia to plunder its nests ; nothing to prevent the full enjoyment of existence. The consequence would be, rapid increase and a large permanent population, which still maintains itself ; for Mr. Moseley, of the *Challenger* expedition, has informed the writer that humming birds are extraordinarily abundant in Juan Fernandez, every bush or tree having one or two darting about it. Here, then, we have one of the special conditions which have always been held to favour variation—a great increase in the number of individuals ; but, as there was no struggle with allied creatures, there was no need for any modification in form or structure, and we accordingly find that the only important variations which have become permanent are those of size and of colour. The increased size would naturally arise from greater abundance of food with a more equable climate throughout the year, the healthier, stronger, and larger individuals being preserved. The change of colour would depend on molecular changes in the plumage accompanying the increase of size ; and the superior energy and vitality in the male, aided by the favourable change in conditions and rapid increase of population, would lead to an increased intensity of colour, the special tint being determined either by local conditions or by inherited tendencies in the race. It is to be noted that the change from green to red is in the direction of the less refrangible rays of the spectrum, and is in accordance with the law of change which has been shown to accompany expansion in inorganic,—growth and development in organic forms.[1] The change of colour in the female, not being urged on by such intense vital activity as in the case of the male, would be much slower, and, owing probably to inherited tendencies, in a different direction. The under-surface of the Chilian bird is ashy with bronzy-green spots on the breast, while the tail is entirely bronze-green. In the Juan Fernandez species the under-surface has become pure white, the breast-spots larger and of a purer golden-green, while the whole inner web of the tail-feathers has become pure white, producing a most elegant effect when the tail is expanded.

We may now follow the two sexes to the remoter

[1] See " Colours of Animals," *Macmillan's Magazine*, Sept. 1877, pp. 394-398, and Chapter V. in the present volume.

island, at a period when the male had acquired his permanent style of colouring, but was not quite so large as he subsequently became ; while the change of the female bird had not been half completed. In this small and comparatively barren island (a mere rock, as it is described by some authors) there would be no such constant abundance of food, and therefore no possibility of a large permanent population; while the climate would not differ materially from that of the larger island. Variation would therefore be checked, or might be stopped altogether ; and we find the facts exactly correspond to this view. The male, which had already acquired his colour, remains almost undistinguishable from his immediate ancestral form ; but he is a little smaller, indicating either that the full size of that form had not been acquired at the period of migration, or that a slight diminution of size has since occurred, owing to a deficiency of food. The female shows also a slight diminution of size, but in other respects is almost exactly intermediate between the Chilian and Juan Fernandez females. The colour beneath is light ashy, the breast-spots are intermediate in size and colour, and the tail-feathers have a large ill-defined white spot on the end of the inner web which has only to be extended along the whole web to produce the exact character which has been acquired in Juan Fernandez. It seems probable, therefore, that the female bird has remained nearly or quite stationary since its migration, while its Juan Fernandez relative has gone on steadily changing in the direction already begun ; and the more distant species geographically thus appears to be more nearly related to its Chilian ancestor.

Coming down to a more recent period, we find that

L

the comparatively small and dull-coloured Chilian bird
has again migrated to Juan Fernandez ; but it at once
came into competition with its red descendant, which
had firm possession of the soil, and had probably undergone slight constitutional changes exactly fitting it to
its insular abode. The new-comer, accordingly, only just
manages to maintain its footing ; for we are told by
Mr. Reed, of Santiago, that it is by no means common;
whereas, as we have seen, the red species is excessively
abundant. We may further suspect that the Chilian
birds now pass over pretty frequently to Juan Fernandez, and thus keep up the stock ; for it must be remembered that whereas, at a first migration, both a male and
a female are necessary for colonization, yet, after a colony
is formed, any stray bird which may come over adds
to the numbers, and checks permanent variation by
cross-breeding.

We find, then, that all the chief peculiarities of the
three allied species of humming-birds which inhabit
the Juan Fernandez group of islands, may be fairly
traced to the action of those general laws which Mr.
Darwin and others have shown to determine the variations of animals and the perpetuation of those variations. It is also instructive to note, that where the
variations of colour and size have been greatest they
are accompanied by several lesser variations in - other
characters. In the Juan Fernandez bird the bill has
become a little shorter, the tail feathers somewhat
broader, and the fiery cap on the head somewhat smaller ;
all these peculiarities being less developed or absent
in the birds inhabiting Mas-afuera. These coincident
changes may be due, either to what Mr. Darwin has

termed correlation of growth, or to the partial reappearance of ancestral characters under more favourable conditions, or to the direct action of changes of climate and of food ; but they show us how varied and unaccountable are the changes in specific forms that may be effected in a comparatively short time, and by means of very slight changes of locality.

If now we consider the enormously varied conditions presented by the whole continent of America—the hot, moist, and uniform forest-plains of the Amazon ; the open llanos of the Orinoco ; the dry uplands of Brazil ; the sheltered valleys and forest slopes of the Eastern Andes ; the verdant plateaus, the barren paramos, the countless volcanic cones with their peculiar Alpine vegetation ; the contrasts of the East and West coasts ; the isolation of the West Indian islands, and to a less extent of Central America and Mexico which we know have been several times separated from South America ; and when we further consider that all these characteristically distinct areas have been subject to cosmical and local changes, to elevations and depressions, to diminution and increase of size, to greater extremes and greater uniformity of temperature, to increase or decrease of rainfall ; and that with these changes there have been coincident changes of vegetation and of animal life, all affecting in countless ways the growth and development, the forms and colours, of these wonderful little birds— if we consider all these varied and complex influences, we shall be less surprised at their strange forms, their infinite variety, their wondrous beauty. For how many ages the causes above enumerated may have acted upon them we cannot say ; but their extreme isolation from all other birds, no less than the abundance and variety of their generic and specific forms, clearly point to a very high antiquity.

The Relations and Affinities of Humming-birds.—The question of the position of this family in the class of birds and its affinities or resemblances to other groups, is so interesting, and affords such good opportunities for explaining some of the best-established principles of classification in natural history in a popular way, that we propose to discuss it at some length, but without entering into technical details.

There is in the Eastern hemisphere, especially in tropical Africa and Asia, a family of small birds called Sun-birds, which are adorned with brilliant metallic colours, and which, in shape and general appearance, much resemble humming-birds. They frequent flowers in the same way, feeding on honey and insects ; and all the older naturalists placed the two families side by side as undoubtedly allied. In the year 1850, in a general catalogue of birds, Prince Lucien Bonaparte, a learned ornithologist, placed the humming-birds next to the swifts, and far removed from the Nectarinidæ or sun-birds ; and this view of their position has gained ground with increasing knowledge, so that now all the more advanced ornithologists have adopted it. Before proceeding to point out the reasons for this change of view, it will be well to discuss a few of the general principles which guide naturalists in the solution of such problems.

How to Determine Doubtful Affinities.—It is now generally admitted that, for the purpose of determining obscure and doubtful affinities, we must examine by

preference those parts of an animal which have little or
no direct influence on its habits and general economy.
The value of an organ, or of any detail of structure,
for purposes of classification, is generally in inverse
proportion to its adaptability to special uses. And the
reason of this is apparent, when we consider that
similarities of food and habits are often accompanied by
similarities of external form or of special organs, in
totally distinct animals. Porpoises, for example, are
modified externally so as to resemble fishes ; yet they
are really mammalia. Some marsupials are carnivorous,
and are so like true carnivora that it is only by minute
peculiarities of structure that the skeleton of the one
can be distinguished from that of the other. Many of
the hornbills and toucans have the same general form,
and resemble each other in habits, in food, and in their
enormous bills ; yet peculiarities in the structure of the
feet, in the form of the breast-bone, in the cranium, and
in the texture and arrangement of the plumage, show
that they have no real affinity, the former approaching
the kingfishers, the latter the cuckoos. Such structural
peculiarities as these have no direct relation to habits ;
and they are therefore little liable to change, when from
any cause a portion of the group may have been driven
to adopt a new mode of life. Thus all the Old World
apes, however much they may differ in size or habits,
and whether we class them as baboons, monkeys, or
gorillas, have the same number of teeth ; while the
American monkeys all have an additional premolar
tooth. This difference can have no relation to the
habits of the two groups, because each group exhibits
differences of habits greater than often occur between

American and Asiatic species ; and it thus becomes a valuable character indicating the radical distinctness of the two groups, a distinctness confirmed by other anatomical characters.

On the other hand, peculiarities of organization which seem specially adapted to certain modes of life, are often diminished or altogether lost in a few species of the group, showing their essential unimportance to the type, as well as their small value for classification. Thus, the woodpeckers are most strikingly characterised by a very long and highly extensible tongue, with the muscles attached to the tongue-bone prolonged backward over the head so as to enable the tongue to be suddenly darted out ; and also by the rigid and pointed tail which is a great help in climbing up the vertical trunks of trees.   But in one group (the Picumni), the tail becomes quite soft, while the tongue remains fully developed ; and in another (Meiglyptes) the characteristic tail remains, while the prolonged hyoid muscles have almost entirely disappeared, and the tongue has consequently lost its peculiar extensile power ; yet in both these cases the form of the breast-bone and the character of the feet, the skeleton, and the plumage, show that the birds are really woodpeckers ; while even the habits and the food are very little altered.   In like manner the bill may undergo great changes ; as from the short crow-like bill of the true birds-of-paradise to the long slender bills of Epimachinæ, which latter were on that account long classed apart in the tribe of Tenuirostres, or slender-billed birds, but whose entire structure shows them to be closely allied to the paradise-birds.   So, the long feathery tongue of the toucans differs from that of every other bird ; yet it is not held to overbalance the weight of anatomical peculiarities which show that these birds are allied to the barbets and the cuckoos.

The skeleton, therefore, and especially the sternum or breast-bone, affords us an almost infallible guide in doubtful cases ; because it appears to change its form with extreme slowness, and thus indicates deeper-seated affinities than those shown by organs which are in direct connection with the outside world, and are readily modified in accordance with varying conditions of existence. Another, though less valuable guide is afforded, in the case of birds, by the eggs. These often have a characteristic form and colour, and a peculiar texture of surface, running unchanged through whole genera and families which are nearly related to each other, however much they may differ in outward form and habits. Another detail of structure which has no direct connection with habits and economy, is the manner in which the plumage is arranged on the body. The feathers of birds are by no means set uniformly over their skin, but grow in certain definite lines and patches, which vary considerably in shape and size in the more important orders and tribes, while the mode of arrangement agrees in all which are known to be closely related to each other ; and thus the form of the feather-tracts or the " pterylography " as it is termed, of a bird, is a valuable aid in doubtful cases of affinity.

Now, if we apply these three tests to the humming-birds, we find them all pointing in the same direction. The sternum or breast-bone is not notched behind ; and this agrees with the swifts, and not with the sun-birds, whose sternum has two deep notches behind, as in all

the families of the vast order of Passeres to which the
latter belong. The eggs of both swifts and humming-
birds are white, only two in number, and resembling
each other in texture. And in the arrangement of the
feather-tracts the humming-birds approach more nearly
to the swifts than they do to any other birds ; and
altogether differ from the sun-birds, which, in this
respect as in so many others, resemble the honey-suckers
of Australia and other true passerine birds.

Resemblances of Swifts and Humming-birds.—Having
this clue to their affinities, we shall find other peculiarities common to these two groups, the swifts and
the humming-birds. They have both ten tail-feathers,
while the sun-birds have twelve. They have both only
sixteen true quill-feathers, and they are the only birds
which have so small a number. The humming-birds
are remarkable for having, in almost all the species,
the first quill the longest of all, the only other birds
resembling them in this respect being a few species
of swifts ; and, lastly, in both groups the plumage
is remarkably compact and closely pressed to the body.
Yet, with all these points of agreement, we find an
extreme diversity in the bills and tongues of the two
groups. The swifts have a short, broad, flat bill, with
a flat horny-tipped tongue of the usual character ; while
the humming-birds have a very long, narrow, almost
cylindrical bill, containing a tubular and highly extensible tongue. The essential point however is, that
whereas hardly any of the other characters we have
adduced are adaptive, or strictly correlated with habits
and economy, this character is pre-eminently so ; for
the swifts are pure aërial insect-hunters, and their short,

broad bills, and wide gape, are essential to their mode
of life. The humming-birds, on the other hand, are
floral insect-hunters, and for this purpose their peculiarly
long bills and extensile tongues are especially adapted ;
while they are at the same time honey-suckers, and
for this purpose have acquired the tubular tongue. The
formation of such a tubular tongue out of one of the
ordinary kind is easily conceivable, as it only requires
to be lengthened, and the two laminæ of which it is
composed curled in at the sides ; and these changes it
probably goes through in the young birds.

When on the Amazon I once had a nest brought me containing two little unfledged humming-birds, apparently
not long hatched. Their beaks were not at all like
those of their parents, but short, triangular, and broad
at the base ; just the form of the beak of a swallow
or swift slightly lengthened. Thinking (erroneously)
that the young birds were fed by their parents on
honey, I tried to feed them with a syrup made of honey
and water, but though they kept their mouths constantly
open as if ravenously hungry, they would not swallow
the liquid, but threw it out again and sometimes nearly
choked themselves in the effort. At length I caught
some minute flies, and on dropping one of these into
the open mouth it instantly closed, the fly was gulped
down and the mouth opened again for more ; and each
took in this way fifteen or twenty little flies in succession
before it was satisfied. They lived thus three or four
days, but required more constant care than I could give
them. These little birds were in the " swift " stage ;
they were pure insect-eaters, with a bill and mouth
adapted for insect-eating only. At that time I was not

aware of the importance of the observation of the tongue ; but as the bill was so short and the tubular tongue not required, there can be little doubt that the organ was, at that early stage of growth, short and flat, as it is in the birds most nearly allied to them.

Differences between Sun-birds and Humming-birds.
—In respect of all the essential and deep-seated points of structure, which have been shown to offer such remarkable similarities between the swifts and the humming-birds, the sun-birds of the Eastern hemisphere differ totally from the latter, while they agree with the passerine birds generally, or more particularly with the creepers and honey-suckers. They have a deeply-notched sternum ; they have twelve tail-feathers in place of ten ; they have nineteen quills in place of sixteen ; and the first quill instead of being the longest is the very shortest of all, while the wings are short and round, instead of being excessively long and pointed ; their plumage is arranged differently ; and their feet are long and strong, instead of being excessively short and weak. There remain only the superficial characters of small size and brilliant metallic colours to assimilate them with the humming-birds, and one structural feature—a tubular and somewhat extensile tongue. This, however, is a strictly adaptive character, the sun-birds feeding on small insects and the nectar of flowers, just as do the humming-birds ; and it is a remarkable instance of a highly peculiar modification of an organ occurring independently in two widely-separate groups. In the sun-birds the hyoid or tongue-muscles do not extend so completely over the head as they do in the humming-birds, so

that the tongue is less extensible; but it is constructed
in exactly the same way by the inrolling of the two
laminæ of which it is composed.

The tubular tongue of the sun-birds is a special
adaptive modification acquired within the family
itself, and not inherited from a remote ancestral form.
This is shown by the amount of variation this organ
exhibits in different members of the family. It is
most highly developed in the Arachnotheræ, or spider-
hunters, of Asia, which are sun-birds without any
metallic or other brilliant colouring. These have the
longest bills and tongues, and the most developed hyoid
muscles; they hunt much about the blossoms of palm-
trees, and may frequently be seen probing the flowers
while fluttering clumsily in the air, just as if they had
seen and attempted to imitate the aërial gambols of
the American humming-birds. The true metallic sun-
birds generally cling about the flowers with their strong
feet; and they feed chiefly on minute hard insects, as
do many humming-birds. There is, however, one species
(Chalcoparia phœnicotis) always classed as a sun-bird,
which differs entirely from the rest of the species in
having the tongue flat, horny, and forked at the tip;
and its food seems to differ correspondingly, for small
caterpillars were found in its stomach. More remotely
allied, but yet belonging to the same family, are the
little flower-peckers of the genus Diceum, which have
a short bill and a tongue twice split at the end; and
these feed on small fruits, and perhaps on buds and on
the pollen of flowers. The little white-eyes (Zosterops),
which are probably allied to the last, eat soft fruits and
minute insects.

Here then we have an extensive group of birds, considerably varied in external form, yet undoubtedly closely allied to each other, one division of which is specially adapted to feed on the juices secreted by flowers and the minute insects that harbour in them ; and these alone have a lengthened bill and double tubular tongue, just as in the humming-birds. We can hardly have a more striking example of the necessity of discriminating between adaptive and purely structural characters. The same adaptive character may coexist in two groups which have a similar mode of life, without indicating any affinity between them, because it may have been acquired by each independently, to enable it to fill a similar place in nature. In such cases it is found to be an almost isolated character, apparently connecting two groups which otherwise differ radically. Non-adaptive, or purely structural characters, on the other hand, are such as have probably been transmitted from a remote ancestor ; and thus indicate fundamental peculiarities of growth and development. The changes of structure rendered necessary by modifications of the habits or instincts of the different species, have been made, to a great extent, independently of such characters ; and as several of these may always be found in the same animal their value becomes cumulative. We thus arrive at the seeming paradox, that the *less* of direct use is apparent in any peculiarity of structure, the *greater* is its value in indicating true, though perhaps remote, affinities ; while any peculiarity of an organ which seems essential to its possessor's well-being is often of very little value in indicating its affinity for other creatures.

This somewhat technical discussion will, it is hoped, enable the general reader to understand some of the more important principles of the modern or natural classification of animals, as distinguished from the artificial system which long prevailed. It will also afford him an easily remembered example of those principles, in the radical distinctness of two families of birds often confounded together,—the sun-birds of the Eastern Hemisphere, and the humming-birds of America; and in the interesting fact that the latter are essentially swifts—profoundly modified, it is true, for an aërial and flower-haunting existence, but still bearing in many important peculiarities of structure the unmistakable evidences of a common origin.

V.

THE COLOURS OF ANIMALS AND SEXUAL SELECTION.

General Phenomena of Colour in the Organic World—Theory of Heat and Light as producing Colour—Changes of Colour in Animals produced by Coloured Light—Classification of Organic Colours—Protective Colours—Warning Colours—Sexual Colours—Typical Colours—The Nature of Colour—Colour a normal product of Organization—Theory of Protective Colours—Theory of Warning Colours—Theory of Sexual Colours—Colour as a means of Recognition—Colour proportionate to Integumentary Development—Selection by Females not a cause of Colour—Probable use of the Horns of Beetles—Cause of the greater brilliancy of some Female Insects—Theory of display of Ornaments by Males—Natural Selection as neutralizing Sexual Selection—Theory of Typical Colours—Colour-development as illustrated by Humming-birds—Local causes of Colour-development—Summary on Colour-development in Animals.

THERE is probably no one quality of natural objects from which we derive so much pure and intellectual enjoyment as from their colours. The heavenly blue of the firmament, the glowing tints of sunset, the exquisite purity of the snowy mountains, and the endless shades of green presented by the verdure-clad surface of the earth, are a never-failing source of pleasure to all who enjoy the inestimable gift of sight. Yet these constitute, as it were, but the frame and background of a marvellous and ever-changing picture. In contrast with these broad and soothing tints, we have presented to us in the vegetable and animal worlds, an infinite variety of objects adorned with the most beautiful and most varied hues. Flowers, insects and birds, are the organisms most generally ornamented in this way; and their symmetry of form, their variety of structure, and the lavish abundance with which they clothe and enliven the earth, cause them to be objects of universal admiration. The relation of this wealth of colour to our mental and moral nature is indisputable. The child and the savage alike admire the gay tints of flower, bird, and insect ; while to many of us their contemplation brings a solace and enjoyment which is both intellectually and morally beneficial. It can then hardly excite surprise that this relation was long thought to afford a sufficient explanation of the phenomena of colour in nature ; and although the fact that—

> " Full many a flower is born to blush unseen,
> And waste its sweetness on the desert air—"

might seem to throw some doubt on the sufficiency of the explanation, the answer was easy,—that in the progress of discovery, man would, sooner or later, find out and enjoy every beauty that the hidden recesses of the earth have in store for him. This theory received great support, from the difficulty of conceiving any other use or meaning in the colours with which so many natural objects are adorned. Why should the homely gorse be clothed in golden raiment, and the prickly cactus be adorned with crimson bells ? Why should our fields be gay with buttercups, and the heather-clad mountains be clad in purple robes ? Why should every land produce its own peculiar floral gems, and the alpine rocks glow with beauty, if not for the contemplation and enjoyment of man ? What could be the use to the butterfly of its

gaily-painted wings, or to the humming bird of its
jewelled breast, except to add the final touches to a world-
picture, calculated at once to please and to refine mankind? And even now, with all our recently-acquired
knowledge of this subject, who shall say that these old-
world views were not intrinsically and fundamentally
sound; and that, although we now know that colour has
" uses " in nature that we little dreamt of, yet the relation
of those colours—or rather of the various rays of light—
to our senses and emotions, may not be another, and
perhaps more important use which they subserve in the
great system of the universe ?

We now propose to lay before our readers a general
account of the more recent discoveries on this interesting
subject; and in doing so, it will be necessary first to
give an outline of the more important facts as to the
colours of organised beings; then to point out the cases
in which it has been shown that colour is of use; and
lastly, to endeavour to throw some light on its nature,
and on the general laws of its development.

Among naturalists, colour was long thought to be of
little import, and to be quite untrustworthy as a specific
character. The numerous cases of variability of colour led
to this view. The occurrence of white blackbirds, white
peacocks, and black leopards; of white blue-bells, and of
white, blue, or pink milkworts, led to the belief that colour
was essentially unstable, that it could therefore be of
little or no importance, and belonged to quite a different
class of characters from form or structure. But it now
begins to be perceived that these cases, though tolerably
numerous, are, after all, exceptional; and that colour, as
a rule, is a constant character. The great majority of

species, both of animals and plants, are each distinguished by peculiar tints which vary very little, while the minutest markings are often constant in thousands or millions of individuals. All our field buttercups are invariably yellow, and our poppies red ; while many of our butterflies and birds resemble each other in every spot and streak of colour through thousands of individuals. We also find that colour is constant in whole genera and other groups of species. The Genistas are all yellow, the Erythrinas all red ; many genera of Carabidæ are entirely black ; whole families of birds—as the Dendrocolaptidæ—are brown ; while among butterflies the numerous species of Lycæna are all more or less blue, those of Pontia white, and those of Callidryas yellow. An extensive survey of the organic world thus leads us to the conclusion that colour is by no means so unimportant or inconstant a character as at first sight it appears to be ; and the more we examine it the more convinced we shall become that it must serve some purpose in nature, and that, besides charming us by its diversity and beauty, it must be well worthy of our attentive study, and have many secrets to unfold to us.

Theory of Heat and Light as producing Colour.—In commencing our study of the great mass of facts relating to the colours of the organic world, it will be necessary to consider first, how far the chief theories already proposed will account for them. One of the most obvious and most popular of these theories, and one which is still held, in part at least, by many eminent naturalists, is—that colour is due to some direct action of the heat and light of the sun—thus at once accounting

M

for the great number of brilliant birds, insects, and flowers, which are found between the tropics.

But before proceeding to discuss this supposed explanation of the colours of living things we must ask the preliminary question,—whether it is really the fact that colour is more developed in tropical than in temperate climates, in proportion to the whole number of species; and even if we find this to be so, we have to inquire whether there are not so many and such striking exceptions to the rule, as to indicate some other causes at work than the direct influence of solar light and heat. As this is a most important branch of the inquiry, we must go into it somewhat fully.

It is undoubtedly the case that there are an immensely greater number of richly-coloured birds and insects in tropical than in temperate and cold countries, but it is by no means so certain that the *proportion* of coloured to obscure species is much or any greater. Naturalists and collectors well know that the majority of tropical birds are dull-coloured; and there are whole families, comprising hundreds of species, not one of which exhibits a particle of bright colour. Such are, for example, the Timaliidæ, or babbling thrushes of the Eastern, and the Dendrocolaptidæ, or tree-creepers of the Western hemispheres. Again, many groups of birds, which are universally distributed, are no more adorned with colour in the tropical than in the temperate zones; such are the thrushes, wrens, goatsuckers, hawks, grouse, plovers, and snipe; and if tropical light and heat have any direct colouring effect, it is certainly most extraordinary that in groups so varied in form, structure, and habits as those just mentioned, the tropical should be in no wise distinguished in this respect, from the temperate species.

It is true that brilliant tropical birds mostly belong to groups which are wholly tropical—as the chatterers, toucans, trogons, and pittas; but as there are perhaps an equal number of groups which are wholly dull-coloured, while others contain dull and bright-coloured species in nearly equal proportions, the evidence is by no means strong that tropical light and heat have anything to do with the matter. But there are other groups in which the cold and temperate zones produce finer-coloured species than the tropics. Thus the arctic ducks and divers are handsomer than those of the tropical zone ; while the king-duck of temperate America and the mandarin-duck of North China are the most beautifully coloured of the whole family. In the pheasant family we have the gorgeous gold and silver pheasants in North China and Mongolia ; and the superb Impeyan pheasant in the temperate North-Western Himalayas, as against the peacock and fire-backed pheasants of tropical Asia. Then we have the curious fact that most of the bright-coloured birds of the tropics are denizens of the forests, where they are shaded from the direct light of the sun, and that they abound near the equator where cloudy skies are very prevalent ; while, on the other hand, places where light and heat are at a maximum have often dull-coloured birds. Such are the Sahara and other deserts, where almost all the living things are sand-coloured ; but the most curious case is that of the Galapagos islands, situated under the equator, and not far from South America where the most gorgeous colours abound, but which are yet characterized by prevailing

dull and sombre tints in birds, insects, and flowers, so
that they reminded Mr. Darwin of the cold and barren
plains of Patagonia rather than of any tropical country.
Insects are wonderfully brilliant in tropical countries
generally ; and any one looking over a collection of South
American or Malayan butterflies would scout the idea of
their being no more gaily-coloured than the average of
European species, and in this he would be undoubtedly
right. But on examination we should find that all the
more brilliantly-coloured groups were exclusively tropical,
and that, where a genus has a wide range, there is little
difference in coloration between the species of cold and
warm countries. Thus the European Vanessides, including the beautiful " peacock," " Camberwell beauty,"
and " red admiral " butterflies, are quite up to the
average of tropical colour in the same group ; and the
remark will equally apply to the little " blues " and
" coppers ; " while the alpine " apollo " butterflies have
a delicate beauty that can hardly be surpassed. In other
insects, which are less directly dependent on climate
and vegetation, we find even greater anomalies. In
the immense family of the Carabidæ or predaceous
ground-beetles, the northern forms fully equal, if they
do not surpass, all that the tropics can produce. Everywhere, too, in hot countries, there are thousands of
obscure species of insects which, if they were all
collected, would not improbably bring down the average
of colour to much about the same level as that of
temperate zones.

But it is when we come to the vegetable world that
the greatest misconception on this subject prevails. In
abundance and variety of floral colour the tropics are

almost universally believed to be pre-eminent, not only absolutely, but relatively to the whole mass of vegetation and the total number of species. Twelve years of observation among the vegetation of the eastern and western tropics has, however, convinced me that this notion is entirely erroneous, and that, in proportion to the whole number of species of plants, those having gaily-coloured flowers are actually more abundant in the temperate zones than between the tropics. This will be found to be not so extravagant an assertion as it may at first appear, if we consider how many of the choicest adornments of our greenhouses and flower-shows are really temperate as opposed to tropical plants. The masses of colour produced by our Rhododendrons, Azaleas, and Camellias, our Pelargoniums, Calceolarias, and Cinerarias,—all strictly temperate plants—can certainly not be surpassed, if they can be equalled, by any productions of the tropics.

It may be objected that most of the plants named are choice cultivated *varieties*, far surpassing in colour the original stock, while the tropical plants are mostly unvaried wild *species*. But this does not really much affect the question at issue. For our florists' gorgeous varieties have all been produced under the influence of our cloudy skies, and with even a still further deficiency of light, owing to the necessity of protecting them under glass from our sudden changes of temperature ; so that they are themselves an additional proof that tropical light and heat are not needed for the production of intense and varied colour. Another important consideration is, that these cultivated *varieties* in many cases displace a number of wild *species* which are hardly, if at all, cultivated. Thus there are scores of *species* of wild hollyhocks varying in colour almost as much as the cultivated varieties, and the same may be said of the pentstemons, rhododendrons, and many other flowers; and if these were all brought together in well-grown specimens, they would produce a grand effect. But it is far easier, and more profitable for our nurserymen to grow *varieties* of one or two species, which all require a similar culture, rather than fifty distinct *species,* most of which would require special treatment; the result being that the varied beauty of the temperate flora is even now hardly known, except to botanists and to a few amateurs.

But we may go further, and say that the hardy plants of our cold temperate zone equal, if they do not surpass, the productions of the tropics. Let us only remember such gorgeous tribes of flowers as the Roses, Pæonies, Hollyhocks, and Antirrhinums; the Laburnum, Wistaria, and Lilac; the Lilies, Irises, and Tulips; the Hyacinths, Anemones, Gentians, and Poppies; and even our humble Gorse, Broom, and Heather; and we may defy any tropical country to produce masses of floral colour in greater abundance and variety. It may be true that individual tropical shrubs and flowers do surpass everything in the rest of the world; but that is to be expected, because the tropical zone comprises a much greater land area than the two temperate zones, while, owing to its more favourable climate, it produces a still larger proportion of species of plants, and a greater number of peculiar natural orders.

Direct observation in tropical forests, plains, and mountains, fully supports this view. Occasionally we are startled by some gorgeous mass of colour, but as a rule we gaze upon an endless expanse of green foliage, only here and there enlivened by not very conspicuous flowers. Even the orchids, whose superb blossoms adorn our stoves, form no exception to this rule. It is only in favoured spots that we find them in abundance ; the species with small and inconspicuous flowers greatly preponderate ; and the flowering season of each kind being of short duration, they rarely produce any marked effect of colour amid the vast masses of foliage which surround them. An experienced collector in the Eastern tropics once told me, that although a single mountain in Java had produced three hundred species of Orchideæ, only about two per cent. of the whole were sufficiently ornamental or showy to be worth sending home as a commercial speculation. The Alpine meadows and rock-slopes, the open plains of the Cape of Good Hope or of Australia, and the flower-prairies of North America, offer an amount and variety of floral colour which can certainly not be surpassed, even if it can be equalled, between the tropics.

It appears, therefore, that we may dismiss the theory that the development of colour in nature is directly dependent on, and in any way proportioned to the amount of solar heat and light, as entirely unsupported by facts. Strange to say, however, there are some rare and little-known phenomena which prove, that in exceptional cases, light does directly affect the colours of natural objects ; and it will be as well to consider these before passing on to other matters.

Changes of Colour in Animals produced by Coloured Light.—A few years ago Mr. T. W. Wood called attention to the curious changes in the colour of the chrysalis of the small cabbage-butterfly (*Pontia rapæ*) when the caterpillars, just before their change, were confined in boxes lined with different tints. Thus in black boxes they were very dark, in white boxes nearly white ; and he further showed that similar changes occurred in a state of nature, chrysalises fixed against a white-washed wall being nearly white ; against a red brick wall, reddish ; against a pitched paling, nearly black. It has also been observed that the cocoon of the emperor moth is either white or brown, according to the colours surrounding it. But the most extraordinary example of this kind of change is that furnished by the chrysalis of an African butterfly (*Papilio Nireus*), observed at the Cape by Mrs. Barber, and described (with a coloured plate) in the *Transactions of the Entomological Society*, 1874, p. 519.

This caterpillar feeds upon the orange tree, and also upon a forest-tree (*Vepris lanceolata*) which has a lighter green leaf ; and its colour corresponds with that of the leaves it feeds upon, being of a darker green when it feeds on the orange. The chrysalis is usually found suspended among the leafy twigs of its food-plant, or of some neighbouring tree, but it is probably often attached to larger branches ; and Mrs. Barber has discovered that it has the property of acquiring the colour, more or less accurately, of any natural object it may be in contact with. A number of the caterpillars were placed in a case with a glass cover, one side of the case being formed by a red brick wall, the other sides being of yellowish wood. They were fed on orange leaves, and a branch of the bottle-brush tree (*Banksia*, *sp.*) was also placed in the case. When fully fed, some attached themselves to the orange twigs, others to the bottle-brush branch; and these all changed to green pupæ; but each corresponded exactly in tint to the leaves around it, the one being dark, the other a pale faded green. Another attached itself to the wood, and the pupa became of the same yellowish colour; while one fixed itself just where the wood and brick joined, and became one side red, the other side yellow! These remarkable changes would perhaps not have been credited had it not been for the previous observations of Mr. Wood; but the two support each other, and oblige us to accept them as actual phenomena. It is a kind of natural photography, the particular coloured rays to which the fresh pupa is exposed in its soft, semi-transparent condition, effecting such a chemical change in the organic juices as to produce the same tint in the hardened skin. It is interesting however to note, that the range of colour that can be acquired seems to be limited to those of natural objects to which the pupa is likely to be attached; for when Mrs. Barber surrounded one of the caterpillars with a piece of scarlet cloth no change of colour at all was produced, the pupa being of the usual green tint, but the small red spots with which it is marked were brighter than usual.

Many other cases are known among insects in which the same species acquires a different tint according to its surroundings; this being particularly marked in some South African locusts, which correspond with the colour of the soil wherever they are found. There are also many caterpillars which feed on two or more plants, and which vary in colour accordingly. A number of such changes are quoted by Mr. R. Meldola, in a paper on Variable Protective Colouring in Insects (*Proceedings of the Zoological Society of London*, 1873, p. 153), and some of them may perhaps be due to a photographic action of the reflected light. In other cases, however, it has been shown that green chlorophyll remains unchanged in the tissues of leaf-eating insects, and being discernible through the transparent integument, produces the same colour as that of the food plant.

In the case of all these insects, as well as in the great majority of cases in which a change of colour occurs in animals, the action is quite involuntary ; but among some of the higher animals the colour of the integument can be modified at the will of the individual, or at all events by a reflex action dependent on sensation. The most remarkable case of this kind occurs with the chameleon, which has the power of changing its colour from dull white to a variety of tints. This singular power has been traced to two layers of movable pigment-cells deeply seated in the skin, but capable of being brought near to the surface. The pigment-layers are bluish and yellowish, and by the pressure of suitable muscles these can be forced upwards either together or separately. When no pressure is exerted the colour is dirty white, which changes to various tints of bluish, green, yellow, or brown, as more or less of either pigment is forced up and rendered visible. The animal is excessively sluggish and defenceless, and its power of changing its colour so as to harmonise with surrounding objects is essential to its safety. Here too, as with the pupa of *Papilio Nireus*, colours, such as scarlet or blue, which do not occur in the immediate environment of the animal, cannot be produced. Somewhat similar changes of colour occur in some prawns and flat-fish, according to the colour of the bottom on which they rest. This is very striking in the chameleon shrimp (*Mysis Chamæleon*), which is grey when on sand, but brown or green when among sea-weed of these two colours. Experiment shows, however, that when blinded the change does not occur ; so that here too we probably have a voluntary or reflex sense-action.

These peculiar powers of change of colour and adaptation are, however, rare and quite exceptional. As a rule, there is no direct connection between the colours of organisms and the kind of light to which they are usually exposed. This is well seen in most fishes and in such marine animals as porpoises, whose backs are always dark, although this part is exposed to the blue and white light of the sky and clouds, while their bellies are very generally white, although these are constantly subjected to the deep blue or dusky green light from the bottom. It is evident, however, that these two tints have been acquired for concealment and protection. Looking *down* on the dark back of a fish it is almost invisible, while, to an enemy looking *up* from below, the light under-surface would be equally invisible against the light of the clouds and sky. Again, the gorgeous colours of the butterflies which inhabit the depths of tropical forests bear no relation to the kind of light that falls upon them, coming as it does almost wholly from green foliage, dark brown soil, or blue sky ; and the bright underwings of many moths, which are only exposed at night, contrast remarkably with the sombre

tints of the upper wings, which are more or less exposed
to the various colours of surrounding nature.

Classification of Organic Colours.—We find, then,
that neither the general influence of solar light and heat,
nor the special action of variously tinted rays, are ade-
quate causes for the wonderful variety, intensity, and
complexity of the colours that everywhere meet us in
the animal and vegetable worlds. Let us therefore take
a wider view of these colours, grouping them into classes
determined by what we know of their actual uses or
special relations to the habits of their possessors. This,
which may be termed the functional and biological classification of the colours of living organisms, seems to
be best expressed by a division into five groups, as
follows :—

	1. Protective colours.	
Animals.	2. Warning colours. { *a.* Of creatures specially protected. *b.* Of defenceless creatures, mimicking *a.*	
	3. Sexual colours.	
	4. Typical colours.	
Plants.	5. Attractive colours.	

It is now proposed, firstly, to point out the nature of
the phenomena presented under each of these heads ;
then to explain the general laws of the production of
colour in nature ; and, lastly, to show how far the varied
phenomena of animal coloration can be explained by
means of those laws, acting in conjunction with the laws
of evolution and natural selection.

Protective Colours.—The nature of the two first
groups, Protective and Warning colours, has been so
fully detailed and illustrated in my chapter on " Mimicry
and other Protective Resemblances among Animals,"
(*Contributions to the Theory of Natural Selection*, p. 45),

that very little need be added here except a few words
of general explanation. Protective colours are exceedingly prevalent in nature, comprising those of all the
white arctic animals, the sandy-coloured desert forms,
and the green birds and insects of tropical forests. It
also comprises thousands of cases of special resemblance
—of birds to the surroundings of their nests, and
especially of insects to the bark, leaves, flowers, or soil,
on or amid which they dwell. Mammalia, fishes, and
reptiles, as well as mollusca and other marine inverte-
brates, present similar phenomena ; and the more the
habits of animals are investigated, the more numerous
are found to be the cases in which their colours tend to
conceal them, either from their enemies or from the
creatures they prey upon. One of the last-observed and
most curious of these protective resemblances has been
communicated to me by Sir Charles Dilke. He was
shown in Java a pink-coloured *Mantis* which, when at
rest, exactly resembled a pink orchis-flower. The
mantis is a carnivorous insect which lies in wait for its
prey ; and, by its resemblance to a flower, the insects it
feeds on would be actually attracted towards it. This
one is said to feed especially on butterflies, so that it
is really a living trap, and forms its own bait !

All who have observed animals, and especially insects,
in their native haunts and attitudes, can understand how
it is that an insect which in a cabinet looks exceedingly
conspicuous, may yet when alive, in its peculiar attitude
of repose and with its habitual surroundings, be perfectly well concealed. We can hardly ever tell by the
mere inspection of an animal, whether its colours are
protective or not. No one would imagine the exquisitely

beautiful caterpillar of the emperor-moth, which is green with pink star-like spots, to be protectively coloured; yet, when feeding on the heather, it so harmonises with the foliage and flowers as to be almost invisible. Every day fresh cases of protective colouring are being discovered, even in our own country; and it is becoming more and more evident that the need of protection has played a very important part in determining the actual coloration of animals.

Warning Colours.—The second class—the warning colours—are exceedingly interesting, because the object and effect of these is, not to conceal the object, but to make it conspicuous. To these creatures it is *useful* to be seen and recognized; the reason being that they have a means of defence which, if known, will prevent their enemies from attacking them, though it is generally not sufficient to save their lives if they are actually attacked. The best examples of these specially protected creatures consist of two extensive families of butterflies, the Danaidæ and Acræidæ, comprising many hundreds of species inhabiting the tropics of all parts of the world. These insects are generally large, are all conspicuously and often most gorgeously coloured, presenting almost every conceivable tint and pattern; they all fly slowly, and they never attempt to conceal themselves; yet no bird, spider, lizard, or monkey (all of which eat other butterflies) ever touches them. The reason simply is that they are not fit to eat, their juices having a powerful odour and taste that is absolutely disgusting to all these animals. Now we see the reason of their showy colours and slow flight. It is good for them to be seen and recognised, for then they are never molested ; but if they did not differ in form and colouring from other butterflies, or if they flew so quickly that their peculiarities could not be easily noticed, they would be captured, and though not eaten would be maimed or killed.

As soon as the cause of the peculiarities of these butterflies was clearly recognised, it was seen that the same explanation applied to many other groups of animals. Thus, bees and wasps and other stinging insects are showily and distinctively coloured ; many soft and apparently defenceless beetles, and many gay-coloured moths, were found to be as nauseous as the above-named butterflies ; other beetles, whose hard and glossy coats of mail render them unpalatable to insect-eating birds, are also sometimes showily coloured ; and the same rule was found to apply to caterpillars, all the brown and green (or protectively coloured species) being greedily eaten by birds, while showy kinds which never hide themselves—like those of the magpie-, mullein-, and burnet-moths—were utterly refused by insectivorous birds, lizards, frogs, and spiders. (*Contributions to the Theory of Natural Selection*, p. 117.) Some few analogous examples are found among vertebrate animals. I will only mention here a very interesting case not given in my former work. In his delightful book entitled, *The Naturalist in Nicaragua*, Mr. Belt tells us that there is in that country a frog which is very abundant ; which hops about in the daytime; which never hides himself ; and which is gorgeously coloured with red and blue. Now frogs are usually green, brown, or earth-coloured ; feed mostly at night ; and are all eaten by snakes and birds. Having full faith in the theory of protective and warning colours, to which he had himself contributed some valuable facts and observations, Mr. Belt felt convinced that this frog must be uneatable. He therefore took one home, and threw it to his ducks and fowls; but all refused to touch it except one young duck, which took the frog in its mouth, but dropped it directly, and went about jerking its head as if trying to get rid of something nasty. Here the uneatableness of the frog was predicted from its colours and habits, and we can have no more convincing proof of the truth of a theory than such previsions.

The universal avoidance by carnivorous animals of all these specially protected groups, which are thus entirely free from the constant persecution suffered by other creatures not so protected, would evidently render it advantageous for any of these latter which were subjected to extreme persecution to be mistaken for the former; and for this purpose it would be necessary that they should have the same colours, form, and habits. Now, strange to say, wherever there is a large group of directly-protected forms (division a of animals with warning colours), there are sure to be found a few otherwise defenceless creatures which resemble them externally so as to be mistaken for them, and which thus gain protection, as it were, on false pretences (division b of animals with warning colours). This is what is called "mimicry," and it has already been very fully treated of by Mr. Bates (its discoverer), by myself, by Mr. Trimen, and others. Here it is only necessary to state that the uneatable Danaidæ and Acræidæ are accompanied by a few species of other groups of butterflies (Leptalidæ, Papilios, Diademas, and Moths) which are all really eatable, but which escape attack by their close resemblance to some species of the uneatable groups found in the same locality. In like manner there are a few eatable beetles which exactly resemble species of uneatable groups ; and others, which are soft, imitate those which are uneatable through their hardness. For the same reason wasps are imitated by moths, and ants by beetles ; and even poisonous snakes are mimicked by harmless snakes, and dangerous hawks by defenceless cuckoos. How these curious imitations have been brought about, and the laws which govern them, have been discussed in the work already referred to.

Sexual Colours.—The third class comprises all cases in which the colours of the two sexes differ. This difference is very general, and varies greatly in amount, from a slight divergence of tint up to a radical change of coloration. Differences of this kind are found among all classes of animals in which the sexes are separated, but they are much more frequent in some groups than in others. In mammalia, reptiles, and fishes, they are comparatively rare, and not great in amount, whereas among birds they are very frequent and very largely developed. So among insects, they are abundant in butterflies, while they are comparatively uncommon in beetles, wasps, and hemiptera.

The phenomena of sexual variations of colour, as well as of colour generally, are wonderfully similar in the two analogous yet totally unrelated groups of birds and butterflies ; and as they both offer ample materials, we shall confine our study of the subject chiefly to them. The most common case of difference of colour between the sexes, is for the male to have the same general hue as the females, but deeper and more

N

intensified; as in many thrushes, finches, and hawks; and among butterflies in the majority of our British species. In cases where the male is smaller the intensification of colour is especially well pronounced; as in many of the hawks and falcons, and in most butterflies and moths in which the coloration does not materially differ. In another extensive series we have spots or patches of vivid colour in the male, which are represented in the female by far less brilliant tints or are altogether wanting; as exemplified in the gold-crest warbler, the green woodpecker, and most of the orange-tip butterflies (*Anthocharis*). Proceeding with our survey, we find greater and greater differences of colour in the sexes, till we arrive at such extreme cases as some of the pheasants, the chatterers, tanagers, and birds-of-paradise, in which the male is adorned with the most gorgeous and vivid colours, while the female is usually dull brown, or olive green, and often shows no approximation whatever to the varied tints of her partner. Similar phenomena occur among butterflies; and in both these groups there are also a considerable number of cases in which both sexes are highly coloured in a different way. Thus many woodpeckers have the head in the male red, in the female yellow; while some parrots have red spots in the male, replaced by blue in the female, as in *Psittacula diopthalma*. In many South American Papilios, green spots on the male are represented by red on the female; and in several species of the genus *Epicalia*, orange bands in the male are replaced by blue in the female, a similar change of colour to that in the small parrot above referred to. For fuller details of the varieties of sexual coloration we refer our readers to Mr. Darwin's *Descent of Man*, chapters x. to xviii., and to chapters iii., iv. and vii. of my *Contributions to the Theory of Natural Selection*.

Typical Colours.—The fourth group—of Typically-coloured animals—includes all species which are brilliantly or conspicuously coloured in both sexes, and for whose particular colours we can assign no function or use. It comprises an immense number of showy birds, such as Kingfishers, Barbets, Toucans, Lories, Tits, and Starlings ; among insects most of the largest and handsomest butterflies, innumerable bright-coloured beetles, locusts, dragon-flies, and hymenoptera ; a few mammalia, as the zebras ; a great number of marine fishes ; thousands of striped and spotted caterpillars ; and abundance of mollusca, star-fish, and other marine animals. Among these we have included some which, like the gaudy caterpillars, have warning colours ; but as that theory does not explain the particular colours or the varied patterns with which they are adorned, it is best to include them also in this class. It is a suggestive fact, that all the brightly-coloured birds mentioned above build in holes or form covered nests, so that the females do not need that protection during the breeding season which I believe to be one of the chief causes of the dull colour of female birds when their partners are gaily coloured. This subject is fully argued in my *Contributions, &c.*, chapter vii.

As the colours of plants and flowers are very different from those of animals both in their distribution and functions, it will be well now to consider how the general facts of colour here sketched out can be explained. We have first to inquire what is colour, and how it is produced ; what is known of the causes of change of colour ; and what theory best accords with the whole assemblage of facts.

The Nature of Colour.—The sensation of colour is caused by vibrations or undulations of the ethereal medium of different lengths and velocities. The whole body of vibrations caused by the sun is termed radiation, or, more commonly, rays ; and consists of sets of waves which vary considerably in their dimensions and rate of recurrence, but of which the middle portion only is capable of exciting in us sensations of light and colour. Beginning with the largest waves, which recur at the longest intervals, we have first those which produce heat-sensations only ; as they get smaller and recur quicker, we perceive a dull red colour ; and as the waves increase in rapidity and diminish in size, we get successively sensations of orange, yellow, green, blue, indigo, and violet, all fading imperceptibly into each other. Then come more invisible rays, of shorter wave-length and quicker recurrence, which produce, solely or chiefly, chemical effects. The red rays, which first become visible, have been ascertained to recur at the rate of 458 millions of millions of times in a second, the length of each wave being $\frac{1}{38900}$th of an inch ; while the violet rays, which last remain visible, recur 727 millions of millions of times per second, and have a wave-length of $\frac{1}{64818}$th of an inch. Although the waves recur at different rates, they are all propagated through the ether with the same velocity (192,000 miles per second) ; just as different musical sounds, which are produced by waves of *air* of different lengths and rates of recurrence, travel at the same speed, so that a tune played several hundred yards off reaches the ear in correct time. There are, therefore, an almost infinite number of different colour-producing undulations, and these may be combined in an almost infinite variety of ways, so as to excite in us the sensation of all the varied colours and tints we are capable of perceiving. When all the different kinds of rays reach us in the proportion in which they exist in the light of the sun, they produce the sensation of white. If the rays which excite the sensation of any one colour are prevented from reaching us, the remaining rays in combination produce a sensation of colour often very far removed from white. Thus green rays being abstracted leave purple light; blue, orange-red light; violet, yellowish-green light, and so on. These pairs are termed complementary colours. And if portions of differently coloured lights are abstracted in various degrees, we have produced all those infinite gradations of colours, and all those varied tints and hues which are of such use to us in distinguishing external objects, and which form one of the great charms of our existence. Primary colours would therefore be as numerous as the different wave-lengths of the visible radiations, if we could appreciate all their differences; while secondary or compound colours, caused by the simultaneous action of any combination of rays of different wave-lengths, must be still more numerous.

In order to account for the fact that all colours appear to us to be produced by combinations of three primary colours—red, green, and violet—it is believed that we have three sets of nerve fibres in the retina, each of which is capable of being excited by all rays, but that one set is excited most by the larger or red waves, another by the medium or green waves, and the third set chiefly by the violet or smallest waves of light ; and when all three sets are excited together in proper proportions we see white. This view is supported by the phenomena of colour-blindness, which are explicable on the theory that one of these sets of nerve-fibres (usually that adapted to perceive red) has lost its sensibility, causing all colours to appear as if the red rays were abstracted from them.

It is a property of these various radiations, that they are unequally refracted or bent in passing obliquely through transparent bodies, the longer waves being least refracted, the shorter most. Hence it becomes possible to analyse white or any other light into its component rays. A small ray of sunlight, for example, which would produce a round white spot on a wall, if passed through a prism is lengthened out into a band of coloured light, exactly corresponding to the colours of the rainbow. Any one colour can thus be isolated and separately examined ; and by means of reflecting mirrors the separate colours can be again compounded in various ways, and the resulting colours observed. This band of coloured light is called a *spectrum*, and the instrument by which the *spectra* of various kinds of light are examined is called a *spectroscope*. This branch of the subject has, however, no direct bearing on the mode in which the colours of living things are produced, and it has only been alluded to in order to complete our sketch of the nature of colour.

The colours which we perceive in material substances are produced either by the absorption or by the interference of some of the rays which form white light. Pigmental or absorption-colours are the most frequent, comprising all the opaque tints of flowers and insects, and all the colours of dyes and pigments. They are caused by rays of certain wave-lengths being absorbed, while the remaining rays are reflected and give rise to the sensation of colour. When all the colour-producing rays are reflected in due proportion, the colour of the object is white ; when all are absorbed the colour is black. If blue rays only are absorbed the resulting colour is orange-red ; and generally, whatever colour an object appears to us, it is because the complementary colours are absorbed by it. The reason why rays of only certain refrangibilities are reflected, and the rest of the incident light absorbed by each substance, is supposed to depend upon the molecular structure of the body. Chemical action almost always implies change of molecular structure, hence chemical action is the most potent cause of change of colour. Sometimes simple solution in water effects a marvellous change, as in the case of the well-known aniline dyes ; the magenta and violet dyes exhibiting, when in the solid form, various shades of golden or bronzy metallic green.

Heat alone often produces change of colour without effecting any chemical change. Mr. Ackroyd has recently investigated this subject,[1] and has shown that a large number of bodies are changed by heat, returning to their normal colour when cooled, and that this change is almost always in the direction of the less refrangible rays or longer wave-lengths ; and he connects the change with the molecular expansion caused by heat.

[1] " Metachromatism, or Colour-Change," *Chemical News*, August, 1876.

As examples may be mentioned mercuric oxide, which is orange yellow, but which changes to orange, red, and brown when heated ; chromic-oxide, which is green, and changes to yellow ; cinnabar, which is scarlet, and changes to puce ; and metaborate of copper, which is blue, and changes to green and greenish yellow.

How Animal Colours are Produced.—The colouring matters of animals are very varied. Copper has been found in the red pigment of the wing of the turaco, and Mr. Sorby has detected no less than seven distinct colouring matters in birds' eggs, several of which are chemically related to those of blood and bile. The same colours are often produced by quite different substances in different groups, as shown by the red of the wing on the burnet-moth changing to yellow with muriatic acid, while the red of the red-admiral-butterfly undergoes no such change.

These pigmental colours have a different character in animals according to their position in the integument. Following Dr. Hagen's classification, epidermal colours are those which exist in the external chitinised skin of insects, in the hairs of mammals, and, partially, in the feathers of birds. They are often very deep and rich, and do not fade after death. The hypodermal colours are those which are situated in the inferior soft layer of the skin. These are often of lighter and more vivid tints, and usually fade after death. Many of the reds and yellows of butterflies and birds belong to this class, as well as the intensely vivid hues of the naked skin about the heads of many birds. These colours sometimes exude through the pores, forming an evanescent bloom on the surface.

Interference colours are less frequent in the organic

world. They are caused in two ways : either by reflection from the two surfaces of transparent films, as seen in the soap-bubble and in thin films of oil on water ; or by fine striæ which produce colours either by reflected or transmitted light, as seen in mother-of-pearl and in finely-ruled metallic surfaces. In both cases colour is produced by light of one wave-length being neutralised, owing to one set of such waves being caused to be half a wave length behind the other set, as may be found explained in any treatise on physical optics. The result is, that the complementary colour of that neutralised is seen ; and, as the thickness of the film or the fineness of the striæ undergo slight changes, almost any colour can be produced. This is believed to be the origin of many of the glossy or metallic tints of insects, as well as those of the feathers of some birds. The iridescent colours of the wings of dragon-flies are caused by the superposition of two or more transparent lamellæ ; while the shining blue of the Purple-Emperor and other butterflies, and the intensely metallic colours of humming-birds, are probably due to fine striæ.

Colour a Normal Product of Organization.—This outline sketch of the nature of colour in the animal world, however imperfect, will at least serve to show us how numerous and varied are the causes which perpetually tend to the production of colour in animal tissues. If we consider, that in order to produce white, all the rays which fall upon an object must be reflected in the same proportions as they exist in solar light—whereas, if rays of any one or more kinds are absorbed or neutralised, the resultant reflected light will be coloured ; and that this colour may be infinitely varied according to the propor-

tions in which different rays are reflected or absorbed—
we should expect that white would be, as it really is,
comparatively rare and exceptional in nature. The same
observation will apply to black, which arises from the
absorption of all the different rays. Many of the
complex substances which exist in animals and plants
are subject to changes of colour under the influence of
light, heat, or chemical change, and we know that
chemical changes are continually occurring during the
physiological processes of development and growth.
We also find that every external character is subject to
minute changes, which are generally perceptible to us
in closely allied species ; and we can therefore have no
doubt that the extension and thickness of the transparent
lamellæ, and the fineness of the striæ or rugosities of
the integuments, must be undergoing constant minute
changes ; and these changes will very frequently produce changes of colour. These considerations render
it probable that colour is a normal and even necessary
result of the complex structure of animals and plants ;
and that those parts of an organism which are undergoing continual development and adaptation to new
conditions, and are also continually subject to the action
of light and heat, will be the parts in which changes of
colour will most frequently appear. Now there is little
doubt that the external changes of animals and plants in
adaptation to the environment are much more numerous
than the internal changes ; as seen in the varied character
of the integuments and appendages of animals—hair,
horns, scales, feathers, &c. &c.—and in plants, the leaves,
bark, flowers, and fruit, with their various modifications
—as compared with the great uniformity in the texture

and composition of their internal tissues ; and this accords with the uniformity of the tints of blood, muscle, nerve, and bone throughout extensive groups, as compared with the great diversity of colour of their external organs. It seems a fair conclusion that colour *per se* may be considered to be normal, and to need no special accounting for ; while the absence of colour (that is, either *white* or *black*), or the prevalence of certain colours to the constant exclusion of others, must be traced, like other modifications in the economy of living things, to the needs of the species. Or, looking at it in another aspect, we may say, that amid the constant variations of animals and plants colour is ever tending to vary and to appear where it is absent; and that natural selection is constantly eliminating such tints as are injurious to the species, or preserving and intensifying such as are useful.

This view is in accordance with the well-known fact, of colours which rarely or never appear in the species in a state of nature, continually occurring among domesticated animals and cultivated plants ; showing us that the capacity to develop colour is ever present, so that almost any required tint can be produced which may, under changed conditions, be useful, in however small a degree.

Let us now see how these principles will enable us to understand and explain the varied phenomena of colour in nature, taking them in the order of our functional classification of colours.

Theory of Protective Colours.—We have seen that obscure or protective tints in their infinitely varied degrees are present in every part of the animal kingdom, whole families or genera being often thus coloured.

Now the various brown, earthy, ashy, and other neutral tints are those which would be most readily produced, because they are due to an irregular mixture of many kinds of rays ; while pure tints require either rays of one kind only, or definite mixtures in proper proportions of two or more kinds of rays. This is well exemplified by the comparative difficulty of producing definite pure tints by the mixture of two or more pigments ; while a haphazard mixture of a number of these will be almost sure to produce browns, olives, or other neutral or dingy colours. An indefinite or irregular absorption of some rays and reflection of others would, therefore, produce obscure tints ; while pure and vivid colours would require a perfectly definite absorption of one portion of the coloured rays, leaving the remainder to produce the true complementary colour. This being the case we may expect these brown tints to occur when the need of protection is very slight or even when it does not exist at all ; always supposing that bright colours are not in any way useful to the species. But whenever a pure colour is protective,—as green in tropical forests or white among arctic snows, there is no difficulty in producing it, by natural selection acting on the innumerable slight variations of tint which are ever occurring. Such variations may, as we have seen, be produced in a great variety of ways ; either by chemical changes in the secretions, or by molecular changes in surface structure ; and may be brought about by change of food, by the photographic action of light, or by the normal process of generative variation. Protective colours therefore, however curious and complex they may be in certain cases, offer no real difficulties.

Theory of Warning Colours.—These differ greatly from the last class, inasmuch as they present us with a variety of brilliant hues, often of the greatest purity, and combined in striking contrasts and conspicuous patterns. Their use depends upon their boldness and visibility, not on the presence of any one colour ; hence we find among these groups some of the most exquisitely-coloured objects in nature. Many of the uneatable caterpillars are strikingly beautiful ; while the Danaidæ, Heliconidæ, and protected groups of Papilionidæ, comprise a series of butterflies of the most brilliant and contrasted colours. The bright colours of many of the sea-anemones and sea-slugs will probably be found to be in this sense protective, serving as a warning of their uneatableness. On our theory none of these colours offer any difficulty. Conspicuousness being useful, every variation tending to brighter and purer colours was selected ; the result being the beautiful variety and contrast we find.

Imitative Warning Colours : — The Theory of Mimicry.—We now come to those groups which gain protection solely by being mistaken for some of these brilliantly coloured but uneatable creatures, and here a difficulty really exists, and to many minds is so great as to be insuperable. It will be well therefore to endeavour to explain how the resemblance in question may have been brought about.

The most difficult case, and the one which may be taken as a type of the whole class, is that of the genus Leptalis (a group of South American butterflies allied to our common white and yellow kinds), many of the larger species of which are still white or yellow, and which are all eatable by birds and other insectivorous creatures. But there are also a number of species of Leptalis, which are brilliantly red, yellow, and black, and which, band for band and spot for spot, resemble some one of the Danaidæ or Heliconidæ which inhabit the same district and which are nauseous and uneatable. Now the usual difficulty is, that a slight approach to one of these protected butterflies would be of no use, while a greater sudden variation is not admissible on the theory of gradual change by indefinite slight variations. This objection depends almost wholly on the supposition that, when the first steps towards mimicry occurred, the South American Danaidæ were what they are now; while the ancestors of the Leptalides were like the ordinary white or yellow Pieridæ to which they are allied. But the danaioid butterflies of South America are so immensely numerous and so greatly varied, not only in colour but in structure, that we may be sure they are of vast antiquity and have undergone great modification. A large number of them, however, are still of comparatively plain colours, often rendered extremely elegant by the delicate transparency of the wing membrane, but otherwise not at all conspicuous. Many have only dusky or purplish bands or spots; others have patches of reddish or yellowish brown—perhaps the commonest colour among butterflies; while a considerable number are tinged or spotted with yellow, also a very common colour, and one especially characteristic of the Pieridæ, the family to which Leptalis belongs. We may therefore reasonably suppose that in the early stages of the development of the Danaidæ, when they first began to acquire those nauseous secretions which are now their protection, their colours were somewhat plain; either dusky with paler bands and spots, or yellowish with dark borders, and sometimes with reddish bands or spots. At this time they had probably shorter wings and a more rapid flight, just like the other unprotected families of butterflies. But as soon as they became decidedly unpalatable to any of their enemies, it would be an advantage to them to be readily distinguished from all the eatable kinds ; and as butterflies were no doubt already very varied in colour, while all probably had wings adapted for rather quick or jerking flight, the best distinction might have been found in outline and habits ; whence would arise the preservation of those varieties whose longer wings, bodies, and antennæ, as well as their slower flight, rendered them noticeable— characters which now distinguish the whole group in every part of the world.

Now it would be at this stage, that some of the weaker-flying Pieridæ which happened to resemble some of the Danaidæ around them in their yellow and dusky tints and in the general outline of their wings, would be sometimes mistaken for them by the common enemy, and would thus gain an advantage in the struggle for existence. Admitting this one step to be made, and all the rest must inevitably follow from simple variation and survival of the fittest. So soon as the nauseous butterfly varied in form or colour to such an extent that the corresponding eatable butterfly no longer closely resembled it, the latter would be exposed to attacks, and only those variations would be preserved which kept up the resemblance. At the same time we may well suppose the enemies to become more acute and able to detect smaller differences than at first. This would lead to the destruction of all adverse variations, and thus keep up in continually increasing complexity the outward mimicry which now so amazes us. During the long ages in which this process has been going on, and the Danaidæ have been acquiring those specialities of colour which aid in their preservation, many a *Leptalis* may have become extinct from not varying sufficiently in the right direction and at the right time to keep up a protective resemblance to its neighbour; and this well accords with the comparatively small number of cases of true mimicry, as compared with the frequency of those protective resemblances to vegetable or inorganic objects whose forms are less definite and colours less changeable. About a dozen other genera of butterflies and moths mimic the Danaidæ in various parts of the world, and exactly the same explanation will apply to all of them. They represent those species of each group which, at the time when the Danaidæ first acquired their protective secretions, happened outwardly to resemble some of them, and which have, by concurrent variation aided by a rigid selection, been able to keep up that resemblance to the present day.[1]

Theory of Sexual Colours.—In Mr. Darwin's celebrated work, *The Descent of Man and Selection in Relation to Sex*, he has treated of sexual colour in combination with other sexual characters, and has

---

[1] For fuller information on this subject the reader should consult Mr. Bates's original paper, "Contributions to an Insect-fauna of the Amazon Valley," in *Transactions of the Linnean Society*, vol. xxiii. p. 495; Mr. Trimen's paper in vol. xxvi. p. 497; the author's essay on "Mimicry," &c., already referred to; and in the absence of collections of butterflies, the plates of Heliconidæ and Leptalidæ, in Hewitson's *Exotic Butterflies*, and Felder's *Voyage of the "Novara,"* may be examined.

arrived at the conclusion that all or almost all the colours of the higher animals (including among these insects and all vertebrates) are due to voluntary or conscious sexual selection; and that diversity of colour in the sexes is due, primarily, to the transmission of colour-variations either to one sex only or to both sexes; the difference depending on some unknown law, and not being due to natural selection.

I have long held this portion of Mr. Darwin's theory to be erroneous; and have argued that the primary cause of sexual diversity of colour was the need of protection, repressing in the female those bright colours which are normally produced in both sexes by general laws; and I have attempted to explain many of the more difficult cases on this principle. ("A Theory of Birds' Nests," in *Contributions, &c.*, p. 231.) As I have since given much thought to this subject, and have arrived at some views which appear to me to be of considerable importance, it will be well to sketch briefly the theory I now hold, and afterwards show its application to some of the detailed cases adduced in Mr. Darwin's work.

The very frequent superiority of the male bird or insect in brightness or intensity of colour, even when the general coloration is the same in both sexes, now seems to me to be, primarily, due to the greater vigour and activity and the higher vitality of the male. The colours of an animal usually fade during disease or weakness, while robust health and vigour adds to their intensity. This is a most important and suggestive fact, and one that appears to hold universally. In all quadrupeds a "dull coat" is indicative of ill-health or low condition; while a glossy coat and sparkling eye are the invariable accompaniments of health and energy. The same rule applies to the feathers of birds, whose colours are only seen in their purity during perfect health; and a similar phenomenon occurs even among insects, for the bright hues of caterpillars begin to fade as soon as they become inactive preparatory to undergoing their transformation. Even in the vegetable kingdom we see the same thing; for the tints of foliage are deepest, and the colours of flowers and fruits richest, on those plants which are in the most healthy and vigorous condition.

This intensity of coloration becomes most developed in the male during the breeding season, when the vitality is at a maximum. It is also very general in those cases in which the male is smaller than the female, as in the hawks and in most butterflies and moths. The same phenomena occur, though in a less marked degree, among mammalia. Whenever there is a difference of colour between the sexes the male is the darker or more strongly marked, and the difference of intensity is most visible during the breeding season (*Descent of Man,* p. 533). Numerous cases among domestic animals also prove, that there is an inherent tendency in the male to special developments of dermal appendages and colour, quite independently of sexual or any other form of selection. Thus,—"the hump on the male zebu cattle of India, the tail of fat-tailed rams, the arched outline of the forehead in the males of several breeds of sheep, and the mane, the long hairs on the hind legs, and the dewlap of the male of the Berbura goat,"—are all adduced by Mr. Darwin as instances of characters peculiar to the male, yet not derived from any parent ancestral form.

Among domestic pigeons the character of the different breeds is often most strongly manifested in the male birds ; the wattles of the carriers and the eye-wattles of the barbs are largest in the males, and male pouters distend their crops to a much greater extent than do the females, while the cock fantails often have a greater number of tail-feathers than the females. There are also some varieties of pigeons of which the males are striped or spotted with black while the females are never so spotted (*Animals and Plants under Domestication*, I. 161) ; yet in the parent stock of these pigeons there are no differences between the sexes either of plumage or colour, and artificial selection has not been applied to produce them.

The greater intensity of coloration in the male— which may be termed the normal sexual difference, would be further developed by the combats of the males for the possession of the females. The most vigorous and energetic usually being able to rear most offspring, intensity of colour, if dependent on, or correlated with vigour, would tend to increase. But as differences of colour depend upon minute chemical or structural differences in the organism, increasing vigour acting unequally on different portions of the integument, and often producing at the same time abnormal developments of hair, horns, scales, feathers, &c., would almost necessarily lead also to variable distribution of colour, and thus to the production of new tints and markings. These acquired colours would, as Mr. Darwin has shown, be transmitted to both sexes or to one only, according as they first appeared at an early age, or in adults of one sex ; and thus we may account for some of the most

marked differences in this respect. With the exception
of butterflies, the sexes are almost alike in the great
majority of insects. The same is the case in mammals
and reptiles ; while the chief departure from the rule
occurs in birds, though even here in very many cases the
law of sexual likeness prevails. But in all cases where
the increasing development of colour became disadvantageous to the female, it would be checked by natural
selection ; and thus produce those numerous instances of
protective colouring in the female only, which occur in
these two groups, birds and butterflies.

Colour as a Means of Recognition.—There is also, I
believe, a very important purpose and use of the varied
colours of the higher animals, in the facility it affords
for recognition by the sexes or by the young of the
same species ; and it is this use which probably fixes
and determines the coloration in many cases. When
differences in size and form are very slight, colour affords
the only means of recognition at a distance, or while in
motion ; and such a distinctive character must therefore
be of especial value to flying insects which are continually in motion, and encounter each other, as it were,
by accident. This view offers us an explanation of the
curious fact, that among butterflies the females of
closely-allied species in the same locality sometimes
differ considerably, while the males are much alike ;
for, as the males are the swiftest and by far the highest
fliers, and seek out the females, it would evidently be
advantageous for them to be able to recognise their true
partners at some distance off. This peculiarity occurs
with many species of *Papilio, Diadema, Adolias,* and
Colias ; and these are all genera, the males of which are

strong on the wing and mount high in the air.   In birds such marked differences of colour are not required, owing to their higher organization and more perfect senses, which render recognition easy by means of a combination of very slight differential characters.

This principle may perhaps, however, account for some anomalies of coloration among the higher animals. Thus, while admitting that the hare and the rabbit are coloured protectively, M r. Darwin remarks that the latter while running to its burrow, is made conspicuous to the sportsman, and no doubt to all beasts of prey, by its upturned white tail.   But this very conspicuousness while running away, may be useful as a signal and guide to the young, who are thus enabled to escape danger by following the older rabbits, directly and without hesitation, to the safety of the burrow ; and this may be the more important from the semi-nocturnal habits of the animal.   If this explanation is correct, and it certainly seems probable, it may serve as a warning of how impossible it is, without exact knowledge of the habits of an animal and a full consideration of all the circumstances, to decide that any particular coloration cannot be protective or in any way useful.   Mr. Darwin himself is not free from such assumptions.   Thus, he says :—" The zebra is conspicuously striped, and stripes cannot afford any protection on the open plains of South Africa." But the zebra is a very swift animal, and, when in herds, by no means void of means of defence.   The stripes therefore *may* be of use by enabling stragglers to distinguish their fellows at a distance, and they *may* be even protective when the animal is at rest among herbage— the only time when it would need protective colouring.

Until the habits of the zebra have been observed with special reference to these points, it is surely somewhat hasty to declare that the stripes "cannot afford any protection."

Colour Proportionate to Integumentary Development. —The wonderful display and endless variety of colour in which butterflies and birds so far exceed all other animals, seems primarily due to the excessive development and endless variations of the integumentary structures. No insects have such widely-expanded wings in proportion to their bodies as butterflies and moths; in none do the wings vary so much in size and form, and in none are they clothed with such a beautiful and highly-organized coating of scales. According to the general principles of the production of colour already explained, these long-continued expansions of membranes and developments of surface structures, must have led to numerous colour-changes; which have been sometimes checked, sometimes fixed and utilised, sometimes intensified, by natural selection, according to the needs of the animal. In birds, too, we have the wonderful clothing of plumage—the most highly organized, the most varied, and the most expanded of all dermal appendages. The endless processes of growth and change during the development of feathers, and the enormous extent of this delicately-organized surface, must have been highly favourable to the production of varied colour-effects; which, when not injurious, have been merely fixed for purposes of specific identification, but have often been modified or suppressed whenever different tints were needed for purposes of protection.

Selection by Females not a Cause of Colour.—To

conscious sexual selection, that is, the actual choice by
the females of the more brilliantly-coloured males, I
believe very little if any effect is directly due. It is
undoubtedly proved that in birds the females do sometimes exert a choice ; but the evidence of this fact
collected by Mr. Darwin (*Descent of Man*, chap. xiv.)
does not prove that colour determines that choice, while
much of the strongest evidence is directly opposed to
this view. All the facts appear to be consistent with
the choice depending on a variety of male characteristics,
with some of which colour is often correlated. Thus it
is the opinion of some of the best observers that vigour
and liveliness are most attractive, and these are no doubt
usually associated with intensity of colour. Again, the
display of the various ornamental appendages of the male
during courtship may be attractive; but these appendages, with their bright colours or shaded patterns, are
due probably to general laws of growth, and to that
superabundant vitality which we have seen to be a
cause of colour. But there are many considerations
which seem to show that the possession of these ornamental appendages and bright colours in the male is not
an important character functionally, and that it has not
been produced by the action of conscious sexual selection.
Amid the copious mass of facts and opinions collected by
Mr. Darwin as to the display of colour and ornaments
by the male birds, there is a total absence of any evidence that the females admire or even notice this display.
The hen, the turkey, and the pea-fowl go on feeding
while the male is displaying his finery ; and there is
reason to believe that it is his persistency and energy
rather than his beauty which wins the day. Again,

evidence collected by Mr. Darwin himself proves that each bird finds a mate under any circumstances. He gives a number of cases of one of a pair of birds being shot, and the survivor being always found paired again almost immediately. This is sufficiently explained on the assumption that the destruction of birds by various causes is continually leaving widows and widowers in nearly equal proportions, and thus each one finds a fresh mate ; and it leads to the conclusion that permanently unpaired birds are very scarce ; so that, speaking broadly, every bird finds a mate and breeds. But this would almost or quite neutralize any effect of sexual selection of colour or ornament, since the less highly-coloured birds would be at no disadvantage as regards leaving healthy offspring. If, however, heightened colour is correlated with health and vigour ; and if these healthy and vigorous birds provide best for their young, and leave offspring which, being equally healthy and vigorous, can best provide for themselves—which cannot be denied ; then natural selection becomes a preserver and intensifier of colour.

Another most important consideration is, that male butterflies rival or even excel the most gorgeous male birds in bright colours and elegant patterns ; and among these there is literally not one particle of evidence that the female is influenced by colour, or even that she has any power of choice; while there is much direct evidence to the contrary (*Descent of Man*, p. 318). The weakness of the evidence for conscious sexual selection among these insects is so palpable, that Mr. Darwin is obliged to supplement it by the singularly inconclusive argument that, "Unless the female prefer one male to another, the pairing must be left to mere chance, and this does not appear probable " (*l. c.* p. 317). But he has just said— " The males sometimes fight together in rivalry, and many may be seen pursuing or crowding round the same female ;" while in the case of the silk-moths,—" the females appear not to evince the least choice in regard to their partners." Surely the plain inference from all this is, that males fight and struggle for the almost passive female; and that the most vigorous and energetic, the strongest-winged or the most persevering, wins her. How can there be chance in this ? Natural selection would here act, as in birds, in perpetuating the strongest and most vigorous males ; and as these would usually be the more highly coloured of their race, the same results would be produced as regards the intensification and variation of colour in the one case as in the other.

Let us now see how these principles will apply to some of the cases adduced by Mr. Darwin in support of his theory of conscious sexual selection.

In *Descent of Man*, 2nd ed., pp. 307-316, we find an elaborate account of the various modes of colouring of butterflies and moths, proving that the coloured parts are always more or less displayed, and that they have some evident relation to an observer. Mr. Darwin then says : " From the several foregoing facts it is impossible to admit that the brilliant colours of butterflies, and of some few moths, have commonly been acquired for the sake of protection. We have seen that their colours and elegant patterns are arranged and exhibited as if for display. Hence I am led to believe that the females prefer or are most excited by the more brilliant males ; for on any other supposition the males would, as far as we can see, be ornamented to no purpose " (*l.c.*, p. 316). I am not aware that any one has ever maintained that the brilliant colours of butterflies have " commonly been acquired for the sake of protection," yet Mr. Darwin has himself referred to cases in which the brilliant colour is so placed as to serve for protection ; as for example, the eye-spots on the hind wings of moths, which are pierced by birds and so save the vital parts of the insect ; while the bright patches on the orange-tip butterflies which Mr. Darwin denies are protective, may serve the same purpose. It is in fact somewhat remarkable how very generally the black spots, ocelli, or bright patches of colour are on the tips, margins, or discs of the wings ; and as the insects are necessarily visible while flying, and this is the time when they are most subject to attacks by insectivorous birds, the position of the more conspicuous parts at some distance from the body may be a real protection to them. Again, Mr. Darwin admits that the white colour of the male ghost-moth may render it more easily seen by the female while flying about in the dusk ; and if to this we add that it will be also more readily distinguished from allied species, we have a reason for diverse ornamentation in these insects quite sufficient to account for most of the facts, without believing in the selection of brilliant males by the females, for which there is not a particle of evidence.[1]

Probable use of the Horns of Beetles.—A somewhat analogous case is furnished by the immense horns of some beetles of the families Copridæ and Dynastidæ, which Mr. Darwin admits are not used for fighting, and

---

[1] See M. Fabre's testimony on this point, *Descent of Man*, p. 291.

therefore concludes are ornaments, developed through
selection of the larger-horned males by the females.
But it has been overlooked that these horns may be
protective. The males probably fly about most, as is
usually the case with male insects; and as they generally fly at dusk they are subject to the attacks of
large-mouthed goatsuckers and podargi, as well as of
insect-eating owls. Now the long, pointed or forked
horns, often divergent, or movable with the head,
would render it very difficult for these birds to swallow
such insects, and would therefore be an efficient protection; just as are the hooked spines of some stingless
ants and the excessively hard integuments of many
beetles, against the smaller insectivorous birds.

*Cause of the greater Brilliancy of some Female In-
sects.*—The facts given by Mr. Darwin to show that
butterflies and other insects can distinguish colours and
are attracted by colours similar to their own, are quite
consistent with the view that colour, which continually
tends to appear, is utilised for purposes of identification and distinction, when not required to be modified
or suppressed for the purpose of protection. The
cases of the females of some species of *Thecla, Callidryas, Colias,* and *Hipparchia,* which have more
conspicuous markings than the male, may be due to
several causes: to obtain greater distinction from other
species; for protection from birds, as in the case of the
yellow-underwing moths; while sometimes—as in *Hipparchia*—the lower intensity of colouring in the female
may lead to more contrasted markings. Mr. Darwin
thinks that here the males have selected the more
beautiful females; although one chief fact in support

of his theory of conscious sexual selection is, that throughout the whole animal kingdom the males are usually so ardent that they will accept any female, while the females are coy, and choose the handsomest males, whence it is believed the general brilliancy of males as compared with females has arisen.

Perhaps the most curious cases of sexual difference of colour are those in which the female is very much more gaily coloured than the male. This occurs most strikingly in some species of *Pieris* in South America, and of *Diadema* in the Malay islands ; and in both cases the females resemble species of the uneatable Danaidæ and Heliconidæ, and thus gain a protection. In the case of *Pieris pyrrha, P. malenka,* and *P. lorena,* the males are plain white and black, while the females are orange, yellow, and black, and so banded and spotted as exactly to resemble species of Heliconidæ. Mr. Darwin admits that these bright colours have been acquired for protection ; but as there is no apparent cause for the strict limitation of the colour to the female, he believes that it has been kept down in the male by its being *unattractive* to her. This appears to me to be a supposition opposed to the whole theory of sexual selection itself. For this theory is, that minute variations of colour in the male are *attractive* to the female, have always been selected, and that thus the brilliant male colours have been produced. But in this case he thinks that the female butterfly had a constant aversion to every trace of colour, even when we must suppose it was constantly recurring during the successive variations which resulted in such a marvellous change in herself. But the case admits of a much more simple interpretation. For if we consider the fact that the females frequent the forests where the Heliconidæ abound, while the males fly much in the open and assemble in great numbers with other white and yellow butterflies on the banks of rivers ; may it not be possible that the appearance of orange stripes or patches would be as injurious to the male as it is useful to the female, by making him a more easy mark for insectivorous birds among his white companions ? This seems a more probable supposition, than the altogether hypothetical choice of the female, sometimes exercised in favour of and sometimes against every new variety of colour in her partner.

A strictly analogous case is that of the glow-worm, whose light, as originally suggested by Mr. Belt, is admitted to be a warning of its uneatability to insectivorous nocturnal animals. The male, having wings, does not require this protection. In the tropics the number of nocturnal insectivorous birds and bats is very much greater, hence winged species possess the light, as they would otherwise be eaten by mistake for more savoury insects ; and it may be that the luminous Elateridæ of the tropics really mimic the true fireflies (Lampyridæ), which are uneatable. This is the more probable as the Elateridæ, in the great majority of species, have brown or protective colours, and are therefore certainly palatable to insectivorous animals.

Origin of the Ornamental Plumage of Male Birds.— We now come to such wonderful developments of plumage and colour as are exhibited by the peacock and the Argus-pheasant ; and I may here mention that it was the case of the latter bird, as fully discussed by

Mr. Darwin, which first shook my belief in "sexual," or more properly "female" selection. The long series of gradations, by which the beautifully shaded ocelli on the secondary wing-feathers of this bird, have been produced, are clearly traced out; the result being a set of markings, so exquisitely shaded as to represent "balls lying loose within sockets"—purely artificial objects of which these birds could have no possible experience. That this result should have been attained through thousands and tens of thousands of female birds all preferring those males whose markings varied slightly in this one direction, this uniformity of choice continuing through thousands and tens of thousands of generations, is to me absolutely incredible. And when, further, we remember that those which did not so vary, would also, according to all the evidence, find mates and leave offspring, the actual result seems quite impossible of attainment by such means.

Without pretending to solve completely so difficult a problem as that of the origin and uses of the variously coloured plumes and ornaments so often possessed by male birds, I would point out a few facts which seem to afford a clue. And first, the most highly-coloured and most richly-varied markings occur on those parts of the plumage which have undergone the greatest modification, or have acquired the most abnormal development. In the peacock, the tail-coverts are enormously developed, and the "eyes" are situated on the greatly dilated ends. In the birds-of-paradise, breast, or neck, or head, or tail-feathers, are greatly developed and highly coloured. The hackles of the cock, and the scaly breasts of humming-birds are similar developments;

while in the Argus-pheasant the secondary quills are so enormously lengthened and broadened as to have become almost useless for flight. Now it is easily conceivable, that during this process of development, inequalities in the distribution of colour may have arisen in different parts of the same feather; and that spots and bands may thus have become broadened out into shaded spots or ocelli, in the way indicated by Mr. Darwin, much as the spots and rings on a soap-bubble increase with increasing tenuity. This is the more probable, because in domestic fowls varieties of colour tend to become symmetrical, quite independently of sexual selection. (*Descent of Man*, p. 424.)

If now we accept the evidence of Mr. Darwin's most trustworthy correspondents, that the choice of the female, so far as she exerts any, falls upon the " most vigorous, defiant, and mettlesome male ; " and if we further believe, what is certainly the case, that these are as a rule the most brightly coloured and adorned with the finest developments of plumage, we have a real and not a hypothetical cause at work. For these most healthy, vigorous, and beautiful males will have the choice of the finest and most healthy females ; will have the most numerous and healthy families ; and will be able best to protect and rear those families. Natural selection, and what may be termed male selection, will tend to give them the advantage in the struggle for existence ; and thus the fullest plumage and the finest colours will be transmitted, and tend to advance in each succeeding generation.

Theory of Display of Ornaments by Males.—The full and interesting account given by Mr. Darwin of the colours and habits of male and female birds (*Descent of Man*, Chapters xiii. and xiv.), proves that in most, if not in all cases, the male birds fully display their ornamental plumage before the females or in rivalry with each other; but on the essential point of whether the female's choice is determined by minute differences in these ornaments or in their colours, there appears to be an entire absence of evidence. In the section on "*Preference for particular Males by the Females*," the facts quoted show indifference to colour, except that some colour similar to their own seems to be preferred. But in the case of the hen canary, who chose a greenfinch in preference to either chaffinch or goldfinch, gay colours had evidently no preponderating attraction. There is some evidence adduced that female birds may, and probably do, choose their mates; but none whatever that the choice is determined by difference of colour; and no less than three eminent breeders informed Mr. Darwin that they "did not believe that the females prefer certain males on account of the beauty of their plumage." Again, Mr. Darwin himself says : "As a general rule colour appears to have little influence on the pairing of pigeons." The oft-quoted case of Sir R. Heron's pea-hens which preferred an "old pied cock" to those normally coloured, is a very unfortunate one; because pied birds are just those that are not favoured in a state of nature, or the breeds of wild animals would become as varied and mottled as our domestic varieties. If such irregular fancies were not rare exceptions, the production of definite colours and patterns by the choice of the female birds, or in any other way, would be impossible.

There remains, however, to be accounted for, the

remarkable fact of the display by the male of each species of its peculiar beauties of plumage and colour,— a display which Mr. Darwin evidently considers his strongest argument in favour of conscious selection by the female. This display is, no doubt, a very interesting and important phenomenon ; but it may, I believe, be satisfactorily explained on the general principles here laid down, without calling to our aid a purely hypothetical choice exerted by the female bird.

At pairing-time, the male is in a state of excitement, and full of exuberant energy. Even unornamental birds flutter their wings or spread them out, erect their tails or crests, and thus give vent to the nervous excitability with which they are overcharged. It is not improbable that crests and other erectile feathers may be primarily of use in frightening away enemies, since they are generally erected when angry or during combat. Those individuals who were most pugnacious and defiant, and who brought these erectile plumes most frequently and most powerfully into action, would tend to increase them by use, and to leave them further developed in some of their descendants. If, in the course of this development, colour appeared—and we have already shown that such developments of plumage are a very probable cause of colour—we have every reason to believe it would be most vivid in these most pugnacious and energetic individuals ; and as these would always have the advantage in the rivalry for mates (to which advantage the excess of colour and plumage might sometimes conduce), there seems nothing to prevent a progressive development of these ornaments in *all dominant races ;* that is, wherever there was such a surplus of vitality, and such

P

complete adaption to conditions, that the inconvenience or danger produced by such ornaments was so comparatively small as not to affect the superiority of the race over its nearest allies.

But if those portions of the plumage, which were originally erected under the influence of anger or fear, became largely developed and brightly coloured, the actual display, under the influence of jealousy or sexual excitement becomes quite intelligible. The males, in their rivalry with each other, would see what plumes were most effective ; and each would endeavour to excel his enemy as far as voluntary exertion would enable him, just as they endeavour to rival each other in song, even sometimes to the point of causing their own destruction.

Natural Selection as Neutralizing Sexual Selection. —There is also a general argument against Mr. Darwin's views on this question, founded on the nature and potency of "natural" as opposed to "sexual" selection, which appears to me to be of itself almost conclusive as to the whole matter at issue. Natural selection, or the survival of the fittest, acts perpetually and on an enormous scale. Taking the offspring of each pair of birds as, on the average, only six annually, one-third of these at most will be preserved, while the two-thirds which are least fitted will die. At intervals of a few years, whenever unfavourable conditions occur, five-sixths, nine-tenths, or even a greater proportion of the whole yearly production are weeded out, leaving only the most perfect and best adapted to survive. Now unless these survivors are, on the whole, the most ornamental, this rigid natural selection must neutralise and destroy any influence that may be exerted by female selection. The utmost that can be claimed for the latter is, that a small fraction of the least ornamented do not obtain mates, while a few of the most ornamented may leave more than the average number of offspring. Unless, therefore, there is the strictest correlation between ornament and general perfection, the more brightly coloured or ornamented varieties can obtain no permanent advantage; and if there is (as I maintain) such a correlation, then the sexual selection of colour or ornament, for which there is little or no evidence, becomes needless, because natural selection. which is an admitted *vera causa*, will itself produce all the results.

In the case of butterflies the argument becomes even stronger, because the fertility is so much greater than in birds, and the weeding-out of the unfit takes place, to a great extent, in the egg and larva state. Unless the eggs and larvæ which escaped to produce the next generation were those which would produce the more highly-coloured butterflies, it is difficult to perceive how the slight preponderance of colour sometimes selected by the females, should not be wholly neutralized by the extremely rigid selection for other qualities to which the offspring in every stage are exposed. The only way in which we can account for the observed facts is, by the supposition that colour and ornament are strictly correlated with health, vigour, and general fitness to survive. We have shown that there is reason to believe that this is the case, and if so, conscious sexual selection becomes as unnecessary as it would certainly be ineffective.

Greater Brilliancy of some Female Birds.—There is one other very curious case of sexual colouring among birds—that, namely, in which the female is decidedly brighter or more strongly marked than the male; as in the fighting quails (*Turnix*), painted snipe (*Rhynchœa*), two species of phalarope (*Phalaropus*), and the common cassowary (*Casuarius galeatus*). In all these cases, it is known that the males take charge of and incubate the eggs, while the females are almost always larger and more pugnacious.

In my "Theory of Birds' Nests" (*Natural Selection*, p. 251), I imputed this difference of colour to the greater need for protection by the male bird while incubating; to which Mr. Darwin has objected that the difference is not sufficient, and is not always so distributed as to be most effective for this purpose; and he believes that it is due to reversed sexual selection, that is, to the female taking the usual *rôle* of the male, and being chosen for her brighter tints. We have already seen reason for rejecting this latter theory in every case; and I also admit that Mr. Darwin's criticism is sound, and that my theory of protection is, in this case, only partially, if at all, applicable. But the theory now advanced, of intensity of colour being due to general vital energy, is quite applicable; and the fact that the superiority of the female in this respect is quite exceptional, and is therefore probably not in any case of very ancient date, will account for the difference of colour thus produced being always very slight.

Colour-development as Illustrated by Humming-birds. —Of the mode of action of the general principles of colour-development among animals, we have an excellent example in the humming-birds. Of all birds these are at once the smallest, the most active, and the fullest of vital energy. When poised in the air their wings are invisible, owing to the rapidity of their motion, and when startled they dart away with the rapidity of a flash of light. Such active creatures would not be an easy prey to any rapacious bird; and if one at length was captured, the morsel obtained would hardly repay the labour. We may be sure, therefore, that they are practically unmolested. The immense variety they exhibit in structure, plumage, and colour, indicates a high antiquity for the race; while their general abundance in individuals shows that they are a dominant group, well adapted to all the conditions of their existence. Here we find everything necessary for the development of colour and accessory plumes. The surplus vital energy shown in their combats and excessive activity, has expended itself in ever-increasing developments of plumage, and greater and greater intensity of colour, regulated only by the need for specific identification which would be especially required in such small and mobile creatures. Thus may be explained those remarkable differences of colour between closely-allied species, one having a crest like the topaz, while in another it resembles the sapphire. The more vivid colours and more developed plumage of the males, I am now inclined to think may be wholly due to their greater vital energy, and to those general laws which lead to such superior developments even in domestic breeds; but in some cases the need of protection by the female while incubating, to which I formerly imputed the whole phenomenon, may have suppressed a portion of the ornament which she would otherwise have attained.

The extreme pugnacity of humming-birds has been

noticed by all observers, and it seems to be to some extent proportioned to the degree of colour and ornament in the species. Thus Mr. Salvin observes of Eugenes fulgens, that it is "a most pugnacious bird," and that "hardly any species shows itself more brilliantly on the wing." Again of Campylopterus hemileucurus,—"the pugnacity of this species is remarkable. It is very seldom that two males meet without an aërial battle,"— and "the large and showy tail of this humming-bird makes it one of the most conspicuous on the wing." Again, the elegant frill-necked Lophornis ornatus " is very pugnacious, erecting its crest, throwing out its whiskers and attacking every humming-bird that may pass within its range of vision ; " and of another species L. magnificus, it is said that "it is so bold that the sight of man creates no alarm." The beautifully-coloured Thaumastura Cora "rarely permits any other humming-bird to remain in its neighbourhood, but wages a continual and terrible war upon them." The magnificent bar-tail, Cometes sparganurus, one of the most imposing of all the humming-birds, is extremely fierce and pugnacious, "the males chasing each other through the air with surprising perseverance and acrimony." These are all the species I find noticed as being especially pugnacious, and every one of them is exceptionally coloured or ornamented ; while not one of the small, plain, and less ornamental species are so described, although many of them are common and well observed species. It is also to be noticed that the remarkable pugnacity of these birds is not confined to one season or even to birds of the same species, as is usual in sexual combats, but extends to any other

species that may be encountered, while they are said
even to attack birds of prey that approach too closely
to their nests. It must be admitted that these facts
agree well with the theory that colour and ornament
are due to surplus vital energy and a long course of
unchecked development. We have also direct evidence
that the males are more active and energetic than the
females. Mr. Gosse says that the whirring made by
the male Polytmus humming-bird is shriller than that
produced by the female ; and he also informs us that
the male flies higher and frequents mountains while the
female keeps to the lowlands.

Theory of Typical Colours.—The remaining kinds of
animal colours, those which can neither be classed as
protective, warning, or sexual, are for the most part
readily explained on the general principles of the development of colour which we have now laid down. It
is a most suggestive fact, that, in cases where colour is
required only as a warning, as among the uneatable
caterpillars, we find, not one or two glaring tints only,
but every kind of colour disposed in elegant patterns,
and exhibiting almost as much variety and beauty as
among insects and birds. Yet here, not only is sexual
selection out of the question, but the need for recognition
and identification by others of the same species, seems
equally unnecessary. We can then only impute this
variety to the normal production of colour in organic
forms, when fully exposed to light and air and undergoing great and rapid developmental modification.
Among more perfect animals, where the need for recognition has been added, we find intensity and variety of
colour at its highest pitch among the South American

butterflies of the families Heliconidæ and Danaidæ, as well as among the Nymphalidæ and Erycinidæ, many of which obtain the necessary protection in other ways. Among birds also, wherever the habits are such that no special protection is needed for the females, and where the species frequent the depths of tropical forests, and are thus naturally protected from the swoop of birds of prey, we find almost equally intense coloration ; as in the trogons, barbets, and gapers.

Local Causes of Colour-development.—Another real, though as yet inexplicable cause of diversity of colour, is to be found in the influence of locality.   It is observed that species of totally distinct groups are coloured alike in one district, while in another district the allied species all undergo the same change of colour. Cases of this kind have been adduced by Mr. Bates, by Mr. Darwin, and by myself, and I have collected all the more curious and important examples in my Address to the Biological Section of the British Association, at Glasgow in 1876 (Chap. VII. of this volume).   The most probable cause for these simultaneous variations would seem to be the presence of peculiar elements or chemical compounds in the soil, the water, or the atmosphere, or of special organic substances in the vegetation ; and a wide field is thus offered for chemical investigation in connection with this interesting subject.   Yet, however we may explain it the fact remains, of the same vivid colours in definite patterns being produced in quite unrelated groups, which only agree, so far as we yet know, in inhabiting the same locality.

Summary on Colour-development in Animals.—Let us now sum up the conclusion at which we have arrived, as to the various modes in which colour is produced or modified in the animal kingdom.

The various causes of colour in the animal world are, molecular and chemical change of the substance of their integuments, or the action on it of heat, light or moisture. It is also produced by interference of light in superposed transparent lamellæ, or by excessively fine surface-striæ. These elementary conditions for the production of colour are found everywhere in the surface-structures of animals, so that its presence must be looked upon as normal, its absence as exceptional.

Colours are fixed or modified in animals by natural selection for various purposes ; obscure or imitative colours for concealment ; gaudy colours as a warning ; and special markings, either for easy recognition by strayed individuals, females, or young, or to direct attack from a vital part, as in the large brilliantly-marked wings of some butterflies and moths.

Colours are produced or intensified by processes of development,—either where the integument or its appendages undergo great extension or modification, or where there is a surplus of vital energy, as in male animals generally, and more especially at the breeding-season.

Colours are also more or less influenced by a variety of causes, such as the nature of the food, the photographic action of light, and also by some unknown local action probably dependent on chemical peculiarities in the soil or vegetation.

These various causes have acted and reacted in a variety of ways, and have been modified by conditions dependent on age or on sex, on competition with new forms, or on geographical or climatic changes. In so complex a subject, for which experiment and systematic inquiry has done so little, we cannot expect to explain every individual case, or solve every difficulty ; but it is believed that all the great features of animal coloration and many of the details become explicable on the principles we have endeavoured to lay down.

It will perhaps be considered presumptuous to put forth this sketch of the subject of colour in animals, as a substitute for one of Mr. Darwin's most highly elaborated theories—that of voluntary or perceptive sexual selection ; yet I venture to think that it is more in accordance with the whole of the facts, and with the theory of natural selection itself ; and I would ask such of my readers as may be sufficiently interested in the subject, to read again Chapters XI. to XVI. of the *Descent of Man*, and consider the whole subject from the point of view here laid down. The explanation of almost all the ornaments and colours of birds and insects as having been produced by the perceptions and choice of the females, has, I believe, staggered many evolutionists, but has been provisionally accepted because it was the only theory that even attempted to explain the facts. It may perhaps be a relief to some of them, as it has been to myself, to find that the phenomena can be shown to depend on the general laws of development, and on the action of "natural selection," which theory will, I venture to think, be relieved from an abnormal excrescence and gain additional vitality, by the adoption of the views here imperfectly set forth.

Although we have arrived at the conclusion that

tropical light and heat can in no sense be considered as the cause of colour, there remains to be explained the undoubted fact that all the more intense and gorgeous tints are manifested by the animal life of the tropics ; while in some groups, such as butterflies and birds, there is a marked preponderance of highly-coloured species. This is probably due to a variety of causes, some of which we can indicate, while others remain to be discovered. The luxuriant vegetation of the tropics throughout the entire year affords so much concealment, that colour may there be safely developed to a much greater extent than in climates where the trees are bare in winter, during which season the struggle for existence is most severe, and even the slightest disadvantage may prove fatal. Equally important, probably, has been the permanence of favourable conditions in the tropics, allowing certain groups to continue dominant for long periods, and thus to carry out in one unbroken line whatever developments of plumage or colour may once have acquired an ascendency. Changes of climatal conditions, and pre-eminently the glacial epoch, probably led to the extinction of a host of highly-developed and finely-coloured insects and birds in temperate zones ; just as we know that it led to the extinction of the larger and more powerful mammalia which formerly characterised the temperate zone in both hemispheres; and this view is supported by the fact that it is amongst those groups only which are now exclusively tropical that all the more extraordinary developments of ornament and colour are found. The obscure local causes of colour to which we have referred will also have acted most efficiently in regions where the climatal condition remained constant, and where migration was unnecessary ; while whatever direct effect may be produced by light or heat, will necessarily have acted more powerfully within the tropics.   And lastly, all these causes have been in action over an actually greater area in tropical than in temperate zones ; while estimated potentially, in proportion to its life-sustaining power, the lands which enjoy a practically tropical climate (extending as they do considerably beyond the geographical tropics) are very much larger than the temperate regions of the earth.

Combining the effects of all these various causes we are quite able to understand the superiority of the tropical parts of the globe, not only in the abundance and variety of their forms of life, but also as regards the ornamental appendages and vivid coloration which these forms present.

VI.

THE COLOURS OF PLANTS AND THE ORIGIN OF THE COLOUR-SENSE.

Source of Colouring matter in Plants—Protective Coloration and Mimicry among Plants—Attractive Colours of Fruits—Protective Colours of Fruits—Attractive Colours of Flowers—Attractive Odours in Flowers—Attractive grouping of Flowers—Why Alpine Flowers are so Beautiful—Why allied species of Flowers differ in Size and Beauty—Absence of Colours in Wind-fertilized Flowers—The same Theory of Colour applicable to Animals and Plants—Relation of the Colours of Flowers and their Geographical Distribution—Recent Views as to the Direct Action of Light on the Colours of Flowers and Fruits—On the Origin of the Colour-sense—Supposed increase of Colour-perception within the Historical Period—Concluding Remarks on the Colour-sense.

THE colouring of plants is neither so varied nor so complex as that of animals, and its explanation accordingly offers fewer difficulties. The colours of foliage are, comparatively, little varied, and can be traced in almost all cases to a special pigment termed chlorophyll, to which is due the general green colour of leaves ; but the recent investigations of Mr. Sorby and others have shown that chlorophyll is not a simple green pigment, but that it really consists of at least seven distinct substances, varying in colour from blue to yellow and orange. These differ in their proportions in the chlorophyll of different plants ; they have different chemical reactions ; they are differently affected by light; and they give distinct spectra. Mr. Sorby further states that scores of different colouring matters are found in the leaves and flowers of plants, to some of which appropriate names have been given, as erythrophyll which is red, and phaiophyll which is brown ; and many of these differ greatly from each other in their chemical composition. These inquiries are at present in their infancy, but as the original term chlorophyll seems scarcely applicable under the present aspect of the subject, it would perhaps be better to introduce the analogous word *Chromophyll*, as a general term for the colouring matters of the vegetable kingdom.

Light has a much more decided action on plants than on animals. The green colour of leaves is almost wholly dependent on it ; and although some flowers will become fully coloured in the dark, others are decidedly affected by the absence of light, even when the foliage is fully exposed to it. Looking therefore at the numerous colouring matters which are developed in the tissues of plants, the sensitiveness of these pigments to light, the changes they undergo during growth and development, and the facility with which new chemical combinations are effected by the physiological processes of plants as shown by the endless variety in the chemical constitution of vegetable products, we have no difficulty in comprehending the general causes which aid in producing the colours of the vegetable world, or the extreme variability of those colours. We may therefore here confine ourselves to an inquiry into the various uses of colour in the economy of plants ; and this will generally enable us to understand how it has become fixed and specialised in the several genera and species of the vegetable kingdom.

Protective Coloration and Mimicry in Plants.—In animals, as we have seen, colour is greatly influenced by the need of protection from, or of warning to, their numerous enemies, and by the necessity for identification and easy recognition. Plants rarely need to be concealed, and obtain protection either by their spines, their hardness, their hairy covering, or their poisonous secretions. A very few cases of what seem to be true protective colouring do, however, exist; the most remarkable being that of the "stone mesembryanthemum," of the Cape of Good Hope, which, in form and colour closely resembles the stones among which it grows; and Dr. Burchell, who first discovered it, believes that the juicy little plant thus generally escapes the notice of cattle and wild herbivorous animals. Mr. J. P. Mansel Weale also noticed that many plants growing in the stony Karoo have their tuberous roots above the soil; and these so perfectly resemble the stones among which they grow that, when not in leaf, it is almost impossible to distinguish them (*Nature*, vol. iii. p. 507). A few cases of what seems to be protective mimicry have also been noted; the most curious being that of three very rare British fungi, found by Mr. Worthington Smith, each in company with common species which they so closely resembled that only a minute examination could detect the difference. One of the common species is stated in botanical works to be "bitter and nauseous," so that it is not improbable that the rare kind may escape being eaten by being mistaken for an uneatable species, though itself palatable. Mr. Mansel Weale also mentions a labiate plant, the *Ajuga-*

ophrydis, of South Africa, as strikingly resembling an orchid. This may be a means of attracting insects to fertilize the flower in the absence of sufficient nectar or other attraction in the flower itself ; and the supposition is rendered more probable by this being the only species of the genus Ajuga in South Africa. Many other cases of resemblances between very distinct plants have been noticed—as that of some Euphorbias to Cacti ; but these very rarely inhabit the same country or locality, and it has not been proved that there is in any of these cases the amount of inter-relation between the species which is the essential feature of the protective " mimicry " that occurs in the animal world.

The different colours exhibited by the foliage of plants and the changes it undergoes during growth and decay, appear to be due to the general laws already sketched out, and to have little if any relation to the special requirements of each species. But flowers and fruits exhibit definite and well-pronounced tints, often varying from species to species, and more or less clearly related to the habits and functions of the plant. With the few exceptions already pointed out, these may be generally classed as *attractive* colours.

Attractive Colours of Fruits.—The seeds of plants require to be dispersed, so as to reach places favourable for germination and growth. Some are very minute, and are carried abroad by the wind ; or they are violently expelled and scattered by the bursting of the containing capsules. Others are downy or winged, and are carried long distances by the gentlest breeze ; or they are hooked and stick to the fur of animals. But there is a large class of seeds which cannot be dispersed in either of these ways, and they are mostly contained in eatable fruits. These fruits are devoured by birds or beasts, and the hard seeds pass through their stomachs undigested, and, owing probably to the gentle heat and moisture to which they have been subjected, in a condition highly favourable for germination. The dry fruits or capsules containing the first two classes of seeds are rarely, if ever, conspicuously coloured; whereas the eatable fruits almost invariably acquire a bright colour as they ripen, while at the same time they become soft and often full of agreeable juices. Our *red* haws and hips, our *black* elderberries, our *blue* sloes, and whortleberries, our *white* mistletoe and snowberry, and our *orange* sea-buckthorn, are examples of the colour-sign of edibility; and in every part of the world the same phenomenon is found. Many such fruits are poisonous to man and to some animals, but they are harmless to others; and there is probably nowhere a brightly-coloured pulpy fruit which does not serve as food for some species of bird or mammal.

Protective Colours of Fruits.—The nuts and other hard fruits of large forest-trees, though often greedily eaten by animals, are not rendered attractive to them by colour, because they are not intended to be eaten. This is evident; for the part eaten in these cases is the seed itself, the destruction of which must certainly be injurious to the species. Mr. Grant Allen, in his ingenious work on *Physiological Æsthetics*, well observes that the colours of all such fruits are protective —green when on the tree, and thus hardly visible among the foliage, but turning brown as they ripen and fall on the ground, as filberts, chestnuts, walnuts, beech-nuts, and many others. It is also to be noted that

Q

many of these are specially though imperfectly protected; some by a prickly coat as in the chestnuts, or by a nauseous covering as in the walnut; and the reason why the protection is not carried further is probably because it is not needed, these trees producing such vast quantities of fruit, that however many are eaten, more than enough are always left to produce young plants. In the case of the attractively coloured fruits, it is curious to observe how the *seeds* are always of such a nature as to escape destruction when the fruit itself is eaten. They are generally very small and comparatively hard, as in the strawberry, gooseberry, and fig; if a little larger, as in the grape, they are still harder and less eatable; in the fruit of the rose (or hip) they are disagreeably hairy; in the orange tribe excessively bitter. When the seeds are larger, softer, and more eatable, they are protected by an excessively hard and stony covering, as in the plum and peach tribe; or they are inclosed in a tough horny core, as with crabs and apples. These last are much eaten by swine, and are probably crushed and swallowed without bruising the core or the seeds, which pass through their bodies undigested. These fruits may also be swallowed by some of the larger frugivorous birds; just as nutmegs are swallowed by pigeons for the sake of the mace which incloses the nut, and which by its brilliant red colour is an attraction as soon as the fruit has split open, which it does upon the tree.

There is, however, one curious case of an attractively coloured seed which has no soft eatable covering. The *Abrus precatoria*, or "rosary bean," is a leguminous shrub or small tree growing in many tropical countries, whose pods curl up and split open on the tree, displaying the brilliant red seeds within. It is very hard and glossy, and is said to be, as no doubt it is, "very indigestible." It may be that birds, attracted by the bright colour of the seeds, swallow them, and that they pass through their bodies undigested, and so get dispersed. If so it would be a case among plants analogous to mimicry among animals—an appearance of edibility put on to deceive birds for the plant's benefit. Perhaps it succeeds only with young and inexperienced birds, and it would have a better chance of success, because such deceptive appearances are very rare among plants.

The smaller plants whose seeds simply drop upon the ground, as in the grasses, sedges, composites, umbelliferæ, &c., always have dry and obscurely-coloured capsules and small brown seeds. Others whose seeds are ejected by the bursting open of their capsules, as with the oxalis and many of the caryophyllaceæ, scrophulariaceæ, &c., have their seeds very small and rarely or never edible.

It is to be remarked that most of the plants whose large seeded nuts cannot be eaten without destroying their germinating power—as the oaks, beeches, and chestnuts—are trees of large size which bear great quantities of fruit, and that they are long lived and have a wide geographical range. They belong to what are called dominant groups, and are thus able to endure having a large proportion of their seeds destroyed with impunity. It is a suggestive fact that they are among the most ancient of known dicotyledonous plants—oaks and beeches going back to the Cretaceous period with little change of type, so that it is not improbable that they may be older than any fruit-eating mammal adapted to feed upon their fruits. The *attractive* coloured fruits on the other hand, having so many special adaptations to dispersal by birds and mammals, are probably of more recent origin.[1] The apple and plum tribes are not known earlier than the Miocene period ; and although the record of extinct vegetable life is extremely imperfect, and the real antiquity of these groups is no doubt very much greater, it is not improbable that the *comparative* antiquity of the fruit-bearing and nut-bearing trees may remain unchanged by further discoveries, as has almost always happened as regards the comparative antiquity of animal groups.

Attractive Colours of Flowers.—The colours of flowers serve to render them visible and recognizable by insects, which are attracted by secretions of nectar or pollen. During their visits for the purpose of obtaining these products, insects involuntarily carry the pollen of one flower to the stigma of another, and thus effect cross-fertilization ; which, as Mr. Darwin was the first to demonstrate, immensely increases the vigour and fertility of the next generation of plants. This discovery has led to the careful examination of great numbers of flowers ; and the result has been that the most wonderful and complex arrangements have been found to exist, all having for their object to secure that flowers shall not be self-fertilized perpetually, but that pollen shall be carried, either constantly or occasionally, from the flowers of one plant to those of another. Mr. Darwin himself first worked out the details in orchids, primulas, and some other groups ; and hardly

[1] I owe this remark to Mr. Grant Allen, author of *Physiological Æsthetics.*

less curious phenomena have since been found to occur
even among some of the most regularly-formed flowers.
The arrangement, length, and position of all the parts
of the flower is now found to have a purpose, and not
the least remarkable portion of the phenomenon is the
great variety of ways in which the same result is
obtained. After the discoveries with regard to orchids,
it was to be expected that the irregular, tubular, and
spurred flowers should present various curious adaptations for fertilization by insect-agency. But even
among the open, cup-shaped, and quite regular flowers,
in which it seemed inevitable that the pollen must fall
on the stigma and produce constant self-fertilization, it
has been found that this is often prevented by a physiological variation—the anthers constantly emitting
their pollen either a little earlier or a little later than
the stigmas of the same flower, or of other flowers on
the same plant, were in the best state to receive it ; and
as individual plants in different stations, soils, and
aspects, differ somewhat in the time of flowering, the
pollen of one plant would often be conveyed by insects
to the stigmas of some other plant in a condition to be
fertilized by it. This mode of securing cross-fertilization
seems so simple and easy, that we can hardly help
wondering why it did not always come into action, and
so obviate the necessity for those elaborate, varied, and
highly complex contrivances found perhaps in the
majority of coloured flowers. The answer to this of
course is, that *variation* sometimes occurred most freely
in one part of a plant's organization, and sometimes
in another ; and that the benefit of cross-fertilization
was so great that *any* variation that favoured it was

preserved, and then formed the starting-point of a whole series of further variations, resulting in those marvellous adaptations for insect fertilization, which have given much of their variety, elegance, and beauty, to the floral world. For details of these adaptations we must refer the reader to the works of Darwin, Lubbock, Herman Müller, and others. We have here only to deal with the part played by colour, and by those floral structures in which colour is most displayed.

Attractive Odours in Flowers.—The sweet odours of flowers, like their colours, seem often to have been developed as an attraction or guide to insect fertilizers, and the two phenomena are often complementary to each other. Thus, many inconspicuous flowers—like the mignonette and the sweet-violet, can be distinguished by their odours before they attract the eye, and this may often prevent their being passed unnoticed; while very showy flowers, and especially those with variegated or spotted petals, are seldom sweet. White, or very pale flowers, on the other hand, are often excessively sweet, as exemplified by the jasmine and clematis; and many of these are only scented at night, as is strikingly the case with the night-smelling stock, our butterfly orchids (*Habenaria chlorantha*), the greenish-yellow *Daphne pontica*, and many others. These white flowers are mostly fertilized by night-flying moths; and those which reserve their odours for the evening probably escape the visits of diurnal insects, which would consume their nectar without effecting fertilization. The absence of odour in showy flowers, and its preponderance among those that are white, may be shown to be a fact by an examination of the lists in Mr. Mongredien's work

on hardy trees and shrubs.[1] He gives a list of about
160 species with showy flowers, and another list of sixty
species with fragrant flowers : but only twenty of these
latter are included among the showy species, and these
are almost all white flowered. Of the sixty species with
fragrant flowers, more than forty are white, and a
number of others have greenish, yellowish, or dusky and
inconspicuous flowers. The relation of white flowers to
nocturnal insects is also well shown by those which,
like the evening primroses, only open their large white
blossoms after sunset. The red Martagon lily has been
observed by Mr. Herman Müller to be fertilized by the
humming-bird hawk moth, which flies in the morning
and afternoon when the colours of this flower, exposed
to the nearly horizontal rays of the sun, glow
with brilliancy, and when it also becomes very sweet-
scented.

Attractive grouping of Flowers.—To the same need of
conspicuousness the combination of so many individually
small flowers into heads and bunches is probably due,
producing such broad masses as those of the elder, the
guelder-rose, and most of the Umbelliferæ, or such
elegant bunches as those of the lilac, laburnum, horse
chestnut, and wistaria. In other cases minute flowers
are gathered into dense heads, as with *Globularia*,
Jasione, clover, and all the Compositæ ; and among the
latter the outer flowers are often developed into a ray, as
in the sunflowers, the daisies, and the asters, forming a
starlike compound flower, which is itself often produced
in immense profusion.

[1] *Trees and Shrubs for English Plantations*, by Augustus Mongredien.
Murray, 1870.

Why Alpine Flowers are so Beautiful.—The beauty
of alpine flowers is almost proverbial. It consists either
in the increased size of the individual flowers as compared with the whole plant, in increased intensity of
colour, or in the massing of small flowers into dense
cushions of bright colour ; and it is only in the higher
Alps, above the limit of forests and upwards towards the
perpetual snow-line that these characteristics are fully
exhibited. This effort at conspicuousness under adverse
circumstances may be traced to the comparative scarcity
of winged insects in the higher regions, and to the
necessity for attracting them from a distance. Amid
the vast slopes of debris and the huge masses of rock so
prevalent in higher mountain regions, patches of intense
colour can alone make themselves visible and serve to
attract the wandering butterfly from the valleys. Mr.
Herman Müller's careful observations have shown, that in
the higher Alps bees and most other groups of winged
insects are almost wanting, while butterflies are tolerably
abundant ; and he has discovered, that in a number of
cases where a lowland flower is adapted to be fertilized by
bees, its alpine ally has had its structure so modified as
to be adapted for fertilization only by butterflies.[1] But
bees are always (in the temperate zone) far more abundant than butterflies, and this will be another reason why
flowers specially adapted to be fertilized by the latter
should be rendered unusually conspicuous. We find,
accordingly, the yellow primrose of the plains replaced by
pink and magenta-coloured alpine species ; the straggling
wild pinks of the lowlands by the masses of large flowers
in such mountain species as *Dianthus alpinus* and *D.*

[1] *Nature*, vol. xi. pp. 32, 110.

glacialis; the saxifrages of the high Alps with bunches of flowers a foot long as in *Saxifraga longifolia* and *S. cotyledon,* or forming spreading masses of flowers as in *S. oppositifolia;* while the soapworts, silenes, and louseworts are equally superior to the allied species of the plains. *Why Allied Species of Flowers Differ in Size and Beauty.*—Again, Dr. Müller has discovered that when there are showy and inconspicuous species in the same genus of plants, there is often a corresponding difference of structure, those with large and showy flowers being quite incapable of self-fertilization, and thus depending for their very existence on the visits of insects ; while the others are able to fertilize themselves should insects fail to visit them. We have examples of this difference in *Malva sylvestris, Epilobium augustifolium, Polygonum bistorta,* and *Geranium pratense*—which have all large or showy flowers, and must be fertilized by insects—as compared with *Malva rotundifolia, Epilobium parviflorum, Polygonum aviculare,* and *Geranium pusillum,* which have small or inconspicuous flowers, and are so constructed that if insects should not visit them they are able to fertilize themselves ?[1]

Absence of Colour in Wind-fertilized Flowers.—As supplementing these curious facts showing the relation of colour in flowers to the need of the visits of insects to fertilize them, we have the remarkable, and on any other theory, utterly inexplicable circumstance, that in all the numerous cases in which plants are fertilized by the agency of the wind they never have specially coloured floral envelopes. Such are our pines, oaks, poplars, willows, beeches, and hazel ; our nettles, grasses,

[1] *Nature,* vol. ix. p. 164.

sedges, and many others. In some of these the male
flowers are, it is true, conspicuous, as in the catkins of
the willows and the hazel, but this arises incidentally
from the masses of pollen necessary to secure fertilization, as shown by the entire absence of a corolla or of
those coloured bracts which so often add to the beauty
and conspicuousness of true flowers.

*The Same Theory of Colour Applicable to Animals
and Plants.*—It may be thought that this absence of
colour where it is not wanted is opposed to the view
maintained in the earlier part of the preceding chapter,
that colour is normal and is constantly tending to appear
in natural objects. It must be remembered, however,
that the green colour of foliage, due to chlorophyll,
prevails throughout the greater part of the vegetable
kingdom, and has, almost certainly, persisted through
long geological periods. It has thus acquired a fixity
of character which cannot be readily disturbed ; and, as
a matter of fact, we find that colour rarely appears in
plants except in association with a considerable modification of leaf-texture, such as occurs in the petals and
coloured sepals of flowers. Wind-fertilized plants never
have such specially organized floral envelopes and, in most
cases, are entirely without a calyx or corolla. The connection between modification of leaf-structure and colour
is further seen in the greater amount and variety of
colour in irregular than in regular flowers. The latter,
which are least modified, have generally uniform or but
slightly varied colours ; while the former, which have
undergone great modification, present an immense range
of colour and marking, culminating in the spotted and
variegated flowers of such groups as the Scrophularineæ

and Orchideæ. The same laws as to the conditions of a maximum production of colour are thus found to obtain both in plants and animals.

Relation of the Colours of Flowers and their Geographical Distribution.—The adaptation of flowers to be fertilized by insects—often to such an extent that the very existence of the species depends upon it—has had wide-spread influence on the distribution of plants and the general aspects of vegetation. The seeds of a particular species may be carried to another country, may find there a suitable soil and climate, may grow and produce flowers ; but if the insect which alone can fertilize it should not inhabit that country, the plant cannot maintain itself, however frequently it may be introduced or however vigorously it may grow. Thus may probably be explained the poverty in flowering-plants and the great preponderance of ferns that distinguishes many oceanic islands, as well as the deficiency of gaily-coloured flowers in others. This branch of the subject is discussed at some length in my Address to the Biological Section of the British Association,[1] but I may here just allude to two of the most striking cases. New Zealand is, in proportion to its total number of flowering-plants, exceedingly poor in handsome flowers, and it is correspondingly poor in insects, especially in bees and butterflies, the two groups which so greatly aid in fertilization. In both these aspects it contrasts strongly with Southern Australia and Tasmania in the same latitudes, where there is a profusion of gaily-coloured flowers and an exceeding rich insect-fauna. The other case is presented by the Galapagos islands, which, though

[1] See Chapter VII. of this volume.

situated on the equator off the west coast of South America, and with a tolerably luxuriant vegetation in the damp mountain zone, yet produce hardly a single conspicuously-coloured flower; and this is correlated with, and no doubt dependent on, an extreme poverty of insect life, not one bee and only a single butterfly having been found there.

Again, there is reason to believe that some portion of the large size and corresponding showiness of tropical flowers is due to their being fertilized by very large insects and even by birds. Tropical sphinx-moths often have their probosces nine or ten inches long, and we find flowers whose tubes or spurs reach about the same length; while the giant bees, and the numerous flower-sucking birds, aid in the fertilization of flowers whose corollas or stamens are proportionately large.

Recent Views as to Direct Action of Light on the Colours of Flowers and Fruits.—The theory that the brilliant colours of flowers and fruits is due to the direct action of light, has been supported by a recent writer by examples taken from the arctic instead of from the tropical flora. In the arctic regions vegetation is excessively rapid during the short summer, and this is held to be due to the continuous action of light throughout the long summer days. "The further we advance towards the north the more the leaves of plants increase in size as if to absorb a greater proportion of the solar rays. M. Grisebach says, that during a journey in Norway he observed that the majority of deciduous trees had already, at the 60th degree of latitude, larger leaves than in Germany, while M. Ch. Martins has made a similar observation as regards the leguminous plants

cultivated in Lapland." [1]   The same writer goes on to
say that all the seeds of cultivated plants acquire a
deeper colour the further north they are grown, white
haricots becoming brown or black, and white wheat
becoming brown, while the green colour of all vegetation
becomes more intense. The flowers also are similarly
changed : those which are white or yellow in central
Europe becoming red or orange in Norway. This is
what occurs in the Alpine flora, and the cause is said
to be the same in both—the greater intensity of the
sunlight.   In the one the light is more persistent, in
the other more intense because it traverses a less thickness of atmosphere.

Admitting the facts as above stated to be in themselves correct, they do not by any means establish the
theory founded on them ; and it is curious that Grisebach,
who has been quoted by this writer for the fact of the
increased size of the foliage, gives a totally different explanation of the more vivid colours of Arctic flowers.
He says—"We see flowers become larger and more
richly coloured in proportion as, by the increasing length
of winter, insects become rarer, and their co-operation
in the act of fecundation is exposed to more uncertain
chances." (*Vegetation du Globe*, vol. i. p. 61 —
French translation.) This is the theory here adopted to
explain the colours of Alpine plants, and we believe
there are many facts that will show it to be the pre-
ferable one.   The statement that the white and yellow
flowers of temperate Europe become red or golden in the
Arctic regions must we think be incorrect.   By roughly

[1] *Revue des Deux Mondes*, 1877.   "La Vegetation dans les hautes Latitudes," par M. Tisserand.

tabulating the colours of the plants given by Sir Joseph Hooker[1] as permanently Arctic, we find among fifty species with more or less conspicuous flowers, twenty-five white, twelve yellow, eight purple or blue, three lilac, and two red or pink ; showing a very similar proportion of white and yellow flowers to what obtains further south.

We have, however, a remarkable flora in the Southern Hemisphere which affords a crucial test of the theory of greater intensity of light being the direct cause of brilliantly coloured flowers. The Auckland and Campbell's Islands south of New Zealand, are in the same latitude as the middle and the south of England, and the summer days are therefore no longer than with us. The climate though cold is very uniform, and the weather "very rainy and stormy." It is evident, then, that there can be no excess of sunshine above what we possess ; yet in a very limited flora there are a number of flowers which—Sir Joseph Hooker states— are equal in brilliancy to the Arctic flora. These con- sist of brilliant gentians, handsome veronicas, large and magnificent Compositæ with purple flowers, bright ranunculi, showy Umbelliferæ, and the golden flowered *Chrysobactron Rossii*, one of the finest of the Aspho- deleæ.[2] All these fine plants, it must be remembered, are peculiar to these islands, and have therefore been developed under the climatal conditions that prevail there ; and as we have no reason to suppose that these conditions have undergone any recent change we may be

---

[1] "On the Distribution of Arctic Plants," *Linn. Trans.* vol. xxiii. (1862.)
[2] Coloured figures of all these plants are given in the *Flora Antarctica*, vol. i.

quite sure that an excess of light has had nothing to do with the development of these exceptionally bright and handsome flowers. Unfortunately we have no information as to the insects of these islands, but from their scarcity in New Zealand we can hardly expect them to be otherwise than very scarce. There are however two species of honey-sucking birds (Prosthemadera and Anthornis) as well as a small warbler (Myiomoira), and we may be pretty sure that the former at least visit these large and handsome flowers, and so effect their fertilization. The most abundant tree on the islands is a species of Metrosideros, and we know that trees of this genus are common in the Pacific islands, where they are almost certainly fertilized by the same family of Meliphagidæ or honey-sucking birds.

I have now concluded this sketch of the general phenomena of colour in the organic world. I have shown reasons for believing that its presence, in some of its infinitely-varied hues, is more probable than its absence; and that variation of colour is an almost necessary concomitant of variation of structure, of development, and of growth. It has also been shown how colour has been appropriated and modified both in the animal and vegetable worlds for the advantage of the species in a great variety of ways, and that there is no need to call in the aid of any other laws than those of organic development and " natural selection " to explain its countless modifications. From the point of view here taken it seems at once improbable and unnecessary that the lower animals should have the same delicate appreciation of the infinite variety and beauty—of the

delicate contrasts and subtle harmonies of colour, which
are possessed by the more intellectual races of mankind,
since even the lower human races do not possess it.   All
that seems required in the case of animals, is a perception of *distinctness* or *contrast* of colours ; and the
dislike of so many creatures to scarlet may perhaps be
due to the rarity of that colour in nature, and to the
glaring contrast it offers to the sober greens and browns
which form the general clothing of the earth's surface,
though it may also have a direct irritating effect on
the retina.

The general view of the subject now given must
convince us that, so far from colour being—as it has
sometimes been thought to be—unimportant, it is intimately connected with the very existence of a large
proportion of the species of the animal and vegetable
worlds.   The gay colours of the butterfly and of the
alpine flower which it unconsciously fertilizes while
seeking for its secreted honey, are each beneficial to its
possessor, and have been shown to be dependent on the
same class of general laws as those which have determined the form, the structure, and the habits of every
living thing.   The complex laws and unexpected
relations which we have seen to be involved in the
production of the special colours of flower, bird, and
insect, must give them an additional interest for every
thoughtful mind ; while the knowledge that, in all
probability, each style of coloration, and sometimes
the smallest details, have a meaning and a use, must
add a new charm to the study of nature.

ON THE ORIGIN OF THE COLOUR-SENSE.

Throughout the preceding discussion we have accepted the subjective phenomena of colour—that is, our perception of varied hues and the mental emotions excited by them, as ultimate facts needing no explanation. Yet they present certain features well worthy of attention, a brief consideration of which will form a fitting sequel to the present essay.

The perception of colour seems, to the present writer, the most wonderful and the most mysterious of our sensations. Its extreme diversities and exquisite beauties seem out of proportion to the causes that are supposed to have produced them, or the physical needs to which they minister. If we look at pure tints of red, green, blue, and yellow, they appear so absolutely contrasted and unlike each other, that it is almost impossible to believe (what we nevertheless know to be the fact) that the rays of light producing these very distinct sensations differ only in wave-length and rate of vibration ; and that there is from one to the other a continuous series and gradation of such vibrating waves. The positive diversity we see in them must then depend upon special adaptations in ourselves ; and the question arises—for what purpose have our visual organs and mental perceptions become so highly specialised in this respect ?

When the sense of sight was first developed in the animal kingdom, we can hardly doubt that what was perceived was light only, and its more or less complete withdrawal. As the sense became perfected, more delicate gradations of light and shade would be perceived ;

and there seems no reason why a visual capacity might
not have been developed as perfect as our own, or even
more so in respect of light and shade, but entirely
insensible to differences of colour except in so far as
these implied a difference in the quantity of light. The
world would in that case appear somewhat as we see it
in good stereoscopic photographs ; and we all know how
exquisitely beautiful such pictures are, and how completely they give us all requisite information as to form,
surface-texture, solidity, and distance, and even to some
extent as to colour ; for almost all colours are dis-
tinguishable in a photograph by some differences of tint,
and it is quite conceivable that visual organs might exist
which would differentiate what we term colour by delicate gradations of some one characteristic neutral tint.
Now such a capacity of vision would be simple as
compared with that which we actually possess ; which,
besides distinguishing infinite gradations of the *quantity*
of light, distinguishes also, by a totally distinct set of
sensations, gradations of *quality*, as determined by
differences of wave-lengths or rate of vibration. At
what grade in animal development this new and more
complex sense first began to appear we have no means
of determining. The fact that the higher vertebrates,
and even some insects, distinguish what are to us
diversities of colour, by no means proves that their
sensations of colour bear any resemblance whatever to
ours. An insect's capacity to distinguish red from blue
or yellow may be (and probably is) due to perceptions
of a totally distinct nature, and quite unaccompanied by
any of that sense of enjoyment or even of radical distinctness which pure colours excite in us. Mammalia

and birds, whose structure and emotions are so similar to our own, do probably receive somewhat similar impressions of colour ; but we have no evidence to show that they experience pleasurable emotions from colour itself, when not associated with the satisfaction of their wants or the gratification of their passions.

The primary necessity which led to the development of the sense of colour, was probably the need of distinguishing objects much alike in form and size, but differing in important properties ;—such as ripe and unripe, or eatable and poisonous fruits ; flowers with honey or without ; the sexes of the same or of closely allied species. In most cases the strongest contrast would be the most useful, especially as the colours of the objects to be distinguished would form but minute spots or points when compared with the broad masses of tint of sky, earth, or foliage against which they would be set.

Throughout the long epochs in which the sense of sight was being gradually developed in the higher animals, their visual organs would be mainly subjected to two groups of rays—the green from vegetation, and the blue from the sky. The immense preponderance of these over all other groups of rays would naturally lead the eye to become specially adapted for their perception ; and it is quite possible that at first these were the only kinds of light-vibrations which could be perceived at all. When the need for differentiation of colour arose, rays of greater and of smaller wave-lengths would necessarily be made use of to excite the new sensations required ; and we can thus understand why green and blue form the central portion of the visible spectrum, and are the colours which are most agreeable to us in large surfaces ;

while at its two extremities we find yellow, red, and violet—colours which we best appreciate in smaller masses, and when contrasted with the other two, or with light neutral tints. We have here probably the foundations of a natural theory of harmonious colouring, derived from the order in which our colour-sensations have arisen and the nature of the emotions with which the several tints have been always associated. The agreeable and soothing influence of green light may be in part due to the green rays having little heating power; but this can hardly be the chief cause, for the blue and violet, though they contain less heat, are not generally felt to be so cool and sedative. But when we consider how dependent are all the higher animals on vegetation, and that man himself has been developed in the closest relation to it, we shall find, probably, a sufficient explanation. The green mantle with which the earth is overspread caused this one colour to predominate over all others that meet our sight, and to be almost always associated with the satisfaction of human wants. Where the grass is greenest, and vegetation most abundant and varied, there has man always found his most suitable dwelling-place. In such spots hunger and thirst are unknown, and the choicest productions of nature gratify the appetite and please the eye. In the greatest heats of summer, coolness, shade, and moisture are found in the green forest glades; and we can thus understand how our visual apparatus has become especially adapted to receive pleasurable and soothing sensations from this class of rays.

Supposed increase of Colour-perception within the Historical Period.—Some writers believe that our power of distinguishing colours has increased even in historical times. The subject has attracted the attention of German philologists, and I have been furnished by a friend with some notes from a work of the late Lazarus Geiger, entitled, *Zur Entwickelungsgeschichte der Menschheit* (Stuttgart, 1871). According to this writer it appears that the *colour* of grass and foliage is never alluded to as a beauty in the Vedas or the Zendavesta, though these productions are continually extolled for other properties. Blue is described by terms denoting sometimes green, sometimes black, showing that it was hardly recognised as a distinct colour. The *colour* of the sky is never mentioned in the Bible, the Vedas, the Homeric poems, or even in the Koran. The first distinct allusion to it known to Geiger is in an Arabic work of the ninth century. "Hyacinthine locks" are black locks, and Homer calls iron "violet-coloured." Yellow was often confounded with green ; but, along with red, it was one of the earliest colours to receive a distinct name. Aristotle names three colours in the rainbow—red, yellow, and green. Two centuries earlier Xenophanes had described the rainbow as purple, reddish, and yellow. The Pythagoreans admitted four primary colours—white, black, red, and yellow.; the Chinese the same, with the addition of green.

Simultaneously with the first publication of this essay in *Macmillan's Magazine,* there appeared in the *Nineteenth Century* an article by Mr. Gladstone on the Colour-sense, chiefly as exhibited in the poems of Homer. He shows that the few colour-terms used by Homer are applied to such different objects that they cannot denote colours only, as we perceive and differentiate them ; but seem more applicable to different intensities of light and shade. Thus, to give one example, the word *porphureos* is applied to clothing, to the rainbow, to blood, to a cloud, to the sea, and to death ; and no one meaning will suit all these applications except comparative darkness. In other cases the same thing has many different epithets applied to it according to its different aspects or conditions ; and as the colours of objects are generally indicated in ancient writings by comparative rather than by abstract terms,—as wine-colour, fire-colour, bronze-colour, &c.—it becomes still more difficult to determine in any particular case what colour was really meant. Mr. Gladstone's general conclusion is, that the archaic man had a positive perception only of degrees of light and darkness, and that in Homer's time he had advanced to the imperfect discrimination of red and yellow, but no further ; the green of grass and foliage or the blue of the sky being never once referred to.

These curious facts cannot, however, be held to prove so recent an origin for colour-sensations as they would at first sight appear to do, because we have seen that both flowers and fruits have become diversely coloured in adaptation to the visual powers of insects, birds, and mammals. Red, being a very common colour of ripe fruits which attract birds to devour them and thus distribute their seeds, we may be sure that the contrast of red and green is to them very well marked. It is indeed just possible that birds may have a more advanced development of the colour-sense than mammals, because the teeth of the latter commonly grind up and destroy the seeds of the larger fruits and nuts which they devour, and which are not usually coloured ; but the irritating effect of bright colours on some of them does not support this view. It seems most probable therefore that man's *perception* of colour in the time of Homer was little if any inferior to what it is now, but that, owing to a variety of causes, no precise *nomenclature* of colours had become established. One of these causes probably was, that the colours of the objects of most importance, and those which were most frequently referred to in songs and poems, were uncertain and subject to variation. Blood was light or dark red, or when dry, blackish ; iron was grey or dark or rusty ; bronze was shining or dull ; foliage was of all shades of yellow, green, or brown ; and horses or cattle had no one distinctive colour. Other objects, as the sea, the sky, and wine, changed in tint according to the light, the time of day, and the mode of viewing them ; and thus colour, indicated at first by reference to certain coloured objects, had no fixity. Things which had more definite and purer colours—as certain species of flowers, birds, and insects—were probably too insignificant or too much despised to serve as colour-terms ; and even these often vary, either in the same or in allied species, in a manner which would render their use unsuitable. Colour-names, being abstractions, must always have been a late development in language, and their comparative unimportance in an early state of society and of the arts would still further retard their appearance ; and this seems quite in accordance with the various facts set forth by Mr. Gladstone and the other writers referred to. The fact that colour-blindness is so prevalent even now, is however an indication that the fully developed colour-sense is not of primary importance to man. If it had been so, natural selection would long ago have eliminated the disease itself, and its tendency to recur would hardly be so strong as it appears to be.

Concluding Remarks on the Colour-sense.—The preceding considerations enable us to comprehend, both why a perception of difference of colour has become developed in the higher animals, and also why colours require to be presented or combined in varying proportions in order to be agreeable to us. But they hardly seem to afford a sufficient explanation, either of the wonderful contrasts and total unlikeness of the sensations produced in us by the chief primary colours, or of the exquisite charm and pleasure we derive from colour itself, as distinguished from variously-coloured objects, in the case of which association of ideas comes into play. It is hardly conceivable that the material *uses* of colour to animals and to ourselves, required such very distinct and powerfully-contrasted sensations ; and it is still less conceivable that a sense of delight in colour *per se* should have been necessary for our utilization of it.

The emotions excited by colour and by music, alike, seem to rise above the level of a world developed on purely utilitarian principles.

VII.

BY-PATHS IN THE DOMAIN OF BIOLOGY :

BEING AN ADDRESS DELIVERED TO THE BIOLOGICAL
SECTION OF THE BRITISH ASSOCIATION, (GLASGOW,
SEPTEMBER 6TH, 1876,) AS PRESIDENT OF THE
SECTION.

Introductory Remarks—On some Relations of Living Things to their
Environment—The Influence of Locality on Colour in Butterflies and
Birds—Sense-perception influenced by Colour of the Integuments—
Relations of Insular Plants and Insects—Rise and Progress of
Modern Views as to the Antiquity and Origin of Man—Indica-
tions of Man's extreme Antiquity—Antiquity of Intellectual Man—
Sculptures on Easter-Island—North American Earthworks—The Great
Pyramid—Conclusion.

THE range of subjects comprehended within the domain
of Biology is so wide, and my own acquaintance with
them so imperfect, that it is not in my power to lay
before you any general outline of the recent progress of
the biological sciences. Neither do I feel competent to
give you a summary of the present status of any one of
the great divisions of our science, such as Anatomy,
Physiology, Embryology, Histology, Classification, or
Evolution—Philology, Ethnology, or Prehistoric Archæo-
logy ; but there are fortunately several outlying and
more or less neglected subjects to which I have for some
time had my attention directed, and which I hope will

furnish matter for a few observations, of some interest to biologists and at the same time not unintelligible to the less scientific members of the Association who may honour us with their presence.

The subjects I first propose to consider have no general name, and are not easily grouped under a single descriptive heading; but they may be compared with that recent development of a sister science which has been termed surface-geology or Earth-sculpture. In the older geological works we learnt much about strata, and rocks, and fossils, their superposition, contortions, chemical constitution, and affinities, with some general notions of how they were formed in the remote past; but we often came to the end of the volume no whit the wiser as to how and why the surface of the earth came to be so wonderfully and beautifully diversified; we were not told why some mountains are rounded and others precipitous; why some valleys are wide and open, others narrow and rocky; why rivers so often pierce through mountain-chains; why mountain-lakes are often so enormously deep; whence came the gravel, and drift, and erratic blocks so strangely spread over wide areas while totally absent from other areas equally extensive. So long as these questions were almost ignored, geology could hardly claim to be a complete science, because, while professing to explain how the crust of the earth came to be what it is, it gave no intelligible account of many phenomena presented by its surface. But of late years these surface-phenomena have been assiduously studied; the marvellous effects of denudation and glacial action in giving the final touches to the actual contour of the earth's surface, and their relation to climatic changes and the antiquity of man, have been clearly traced, thus investing geology with a new and popular interest, and at the same time elucidating many of the phenomena presented in the older formations.

Now just as a surface-geology was required to complete that science, so a surface-biology was wanted to make the science of living things more complete and more generally interesting, by applying the results arrived at by special workers to the interpretation of those external and prominent features whose endless variety and beauty constitute the charm which attracts us to the contemplation or to the study of nature. We have the descriptive zoologist, for example, who gives us the external characters of animals ; the anatomist studies their internal structure ; the histologist makes known the nature of their component tissues ; the embryologist patiently watches the progress of their development ; the systematist groups them into classes and orders, families, genera, and species ; while the field-naturalist studies for us their food and habits and general economy. But, till quite recently, none of these earnest students nor all of them combined, could answer satisfactorily, or even attempted to answer, many of the simplest questions concerning the external characters and general relations of animals and plants. Why are flowers so wonderfully varied in form and colour ? what causes the Arctic fox and the ptarmigan to turn white in winter ? why are there no elephants in America and no deer in Australia ? why are closely allied species rarely found together ? why are male animals so frequently bright-coloured ? why are extinct animals so often larger than those which are now living ? what has led to the production of the

gorgeous train of the peacock and of the two kinds of
flower in the primrose ?   The solution of these and a
hundred other problems of like nature was rarely approached by the old method of study, or if approached
was only the subject of vague speculation.   It is to the
illustrious author of the *Origin of Species* that we
are indebted for teaching us how to study nature as one
great, compact, and beautifully-adjusted system.   Under
the touch of his magic wand the countless isolated facts
of internal and external structure of living things—
their habits, their colours, their development, their distribution, their geological history,—all fell into their ap-
proximate places ; and although, from the intricacy of
the subject and our very imperfect knowledge of the
facts themselves, much still remains uncertain, yet we
can no longer doubt that even the minutest and most
superficial peculiarities of animals and plants either, on
the one hand, are or have been useful to them, or, on
the other hand, have been developed under the influence
of general laws, which we may one day understand to a
much greater extent than we do at present.   So great is
the alteration effected in our comprehension of nature
by the study of variation, inheritance, cross-breeding,
competition, distribution, protection, and selection—
showing, as they often do, the meaning of the most
obscure phenomena and the mutual dependence of the
most widely-separated organisms—that it can only be
fitly compared with the analogous alteration produced
in our conception of the universe by Newton's grand
discovery of the law of gravitation.

I know it will be said (and is said), that Darwin is
too highly rated, that some of his theories are wholly

and others partially erroneous, and that he often builds a vast superstructure on a very uncertain basis of doubtfully interpreted facts. Now, even admitting this criticism to be well founded—and I myself believe that to a limited extent it is so—I nevertheless maintain that Darwin is not and cannot be too highly rated ; for his greatness does not at all depend upon his being infallible, but on his having developed, with rare patience and judgment, a new system of observation and study, guided by certain general principles which are almost as simple as gravitation and as wide-reaching in their effects. And if other principles should hereafter be discovered, or if it be proved that some of his subsidiary theories are wholly or partially erroneous, this very discovery can only be made by following in Darwin's steps, by adopting the method of research which he has taught us, and by largely using the rich stores of material which he has collected. The *Origin of Species*, and the grand series of works which have succeeded it, have revolutionized the study of biology ; they have given us new ideas and fertile principles ; they have infused life and vigour into our science, and have opened up hitherto unthought-of lines of research on which hundreds of eager students are now labouring. Whatever modifications some of his theories may require, Darwin must none the less be looked up to as the founder of philosophical biology.

As a small contribution to this great subject, I propose now to call your attention to some curious relations of organisms to their environment, which seem to me worthy of more systematic study than has hitherto been given them. The points I shall more especially deal

with are—the influence of locality, or of some unknown
local causes, in determining the colours of insects, and,
to a less extent, of birds ; and the way in which certain
peculiarities in the distribution of plants may have been
brought about by their dependence on insects. The
latter part of my address will deal with the present state
of our knowledge as to the antiquity and early history
of mankind.

ON SOME RELATIONS OF LIVING THINGS TO THEIR ENVIRONMENT.

Of all the external characters of animals, the most
beautiful, the most varied, and the most generally
attractive are the brilliant colours and strange yet often
elegant markings with which so many of them are
adorned.   Yet of all characters this is the most difficult
to bring under the laws of utility or of physical connection.   Mr. Darwin—as you are well aware—has
shown how wide is the influence of sex on the intensity
of coloration ; and he has been led to the conclusion
that active or voluntary sexual selection is one of the
chief causes, if not the chief cause, of all the variety
and beauty of colour we see among the higher animals.
This is one of the points on which there is much divergence of opinion even among the supporters of Mr.
Darwin, and one as to which I myself differ from him.
I have argued, and still believe, that the need of protection is a far more efficient cause of variation of colour
than is generally suspected ; but there are evidently
other causes at work, and one of these seems to be an
influence depending strictly on locality, whose nature

we cannot yet understand, but whose effects are everywhere to be seen when carefully searched for.

Although the careful experiments of Sir John Lubbock have shown that insects can distinguish colours—as might have been inferred from the brilliant colours of the flowers which are such an attraction to them—yet we can hardly believe that their appreciation and love of distinctive colours is so refined as to guide and regulate their most powerful instinct—that of reproduction. We are therefore led to seek some other cause for the varied colours that prevail among insects ; and as this variety is most conspicuous among butterflies—a group perhaps better known than any other—it offers the best means of studying the subject. The variety of colour and marking among these insects is something marvellous. There are probably about ten thousand different kinds of butterflies now known, and about half of these are so distinct in colour and marking that they can be readily distinguished by this means alone. Almost every conceivable tint and pattern is represented, and the hues are often of such intense brilliance and purity as can be equalled by neither birds nor flowers.

Any help to a comprehension of the causes which may have concurred in bringing about so much diversity and beauty must be of value ; and this is my excuse for laying before you the more important cases I have met with of a connection between colour and locality.

The influence of Locality on Colour in Butterflies and Birds.—Our first example is from tropical Africa, where we find two unrelated groups of butterflies belonging to two very distinct families (Nymphalidæ and Papilionidæ) characterized by a prevailing blue-

green colour not found in any other continent.[1] Again,
we have a group of African Pieridæ which are white
or pale yellow with a marginal row of bead-like black
spots ; and in the same country one of the Lycænidæ
(*Leptena erastus*) is coloured so exactly like these that
it was at first described as a species of *Pieris*. None of
these four groups are known to be in any way specially
protected, so that the resemblance cannot be due to
protective mimicry.

In South America we have far more striking cases ;
for in the three subfamilies Danainæ, Acræinæ, and
Heliconiinæ, all of which are specially protected, we
find identical tints and patterns reproduced, often in the
greatest detail, each peculiar type of coloration being
characteristic of separate geographical subdivisions of
the continent.   Nine very distinct genera are implicated
in these parallel changes—Lycorea, Ceratinia, Mechanitis, Ithomia, Melinæa, Tithorea, Acræa, Heliconius,
and Eueides, groups of three or four (or even five) of
them appearing together in the same livery in one
district, while in an adjoining district most or all of
them undergo a simultaneous change of coloration or of
marking.   Thus in the genera Ithomia, Mechanitis,
and Heliconius we have species with yellow apical spots
in Guiana, all represented by allied species with white
apical spots in South Brazil.   In Mechanitis, Melinæa,
and Heliconius, and sometimes in Tithorea, the species
of the Southern Andes (Bolivia and Peru) are characterized by an orange and black livery, while those of
the Northern Andes (New Granada) are almost always

---

[1] Romaleosoma and Euryphene (Nymphalidæ), *Papilio zalmoxis* and
several species of the Nireus-group (Papilionidæ).

orange-yellow and black.   Other changes of a like
nature, which it would be tedious to enumerate but
which are very striking when specimens are examined,
occur in species of the same groups inhabiting these
same localities, as well as Central America and the
Antilles. The resemblance thus produced between widely
different insects is sometimes general, but often so close
and minute that only a critical examination of structure
can detect the difference between them.   Yet this can
hardly be true mimicry, because all are alike protected
by the nauseous secretion which renders them unpalatable to birds.

In another series of genera (Catagramma, Callithea,
and Agrias) all belonging to the Nymphalidæ, we have
the most vivid blue ground, with broad bands of orange,
crimson or a different tint of blue or purple, exactly
reproduced in corresponding, yet unrelated species,
occurring in the same locality ; yet, as none of these
groups are known to be specially protected, this can
hardly be true mimicry.   A few species of two other
genera in the same country (Eunica and Siderone) also
reproduce the same colours, but with only a general
resemblance in the markings.   Yet again, in tropical
America we have species of Apatura which, sometimes
in both sexes, sometimes in the female only, exactly
imitate the peculiar markings of another genus (Heterochroa) confined to America : here, again, neither genus
is protected, and the similarity must be due to unknown
local causes.

But it is among islands that we find some of the most
striking examples of the influence of locality on colour,
generally in the direction of paler, but sometimes of

s

darker and more brilliant hues, and often accompanied by an unusual increase of size. Thus in the Moluccas and New Guinea we have several Papilios (*P. euchenor*, *P. ormenus*, and *P. tydeus*) distinguished from their allies by a much paler colour, especially in the females which are almost white. Many species of Danais (forming the subgenus Ideopsis) are also very pale. But the most curious are the Euplœas, which in the larger islands are usually of rich dark colours, while in the small islands of Banda, Ké, and Matabello at least three species not nearly related to each other (*E. hoppferi*, *E. euripon*, and *E. assimilata*) are all broadly banded or suffused with white, their allies in the larger islands being all very much darker. Again, in the genus Diadema, belonging to a distinct family, three species from the small Aru and Ké islands (*D. deois*, *D. hewitsonii*, and *D. polymena*) are all more conspicuously white-marked than their representatives in the larger islands. In the beautiful genus Cethosia, a species from the small island of Waigiou (*C. cyrene*) is the whitest of the genus. Prothoë is represented by a blue species in the continental island of Java, while those inhabiting the ancient insular groups of the Moluccas and New Guinea are all pale yellow or white. The genus Drusilla, almost confined to these islands, comprises many species which are all very pale ; while in the small island of Waigiou is found a very distinct genus, Hyantis, which, though differing completely in the neuration of the wings, has exactly the same pale colours and large ocellated spots as Drusilla.

Equally remarkable is the increase of size in some islands. The small island of Amboina produces larger butterflies than any of the much larger islands which surround it. This is the case with at least a dozen butterflies belonging to many distinct genera,[1] so that it is impossible to attribute the fact to other than some local influence. In Celebes, as I have elsewhere pointed out,[2] we have a peculiar form of wing and much larger size running through a whole series of distinct butterflies ; and this seems to take the place of any speciality in colour.

In a very small collection of insects recently brought from Duke-of-York Island (situated between New Britain and New Ireland) are several of remarkably white or pale coloration. A species of Euplæa is the whitest of all known species of that extensive genus ; while a beautiful diurnal moth is much whiter than its ally in the larger island of New Guinea. There is also a magnificent longicorn beetle almost entirely of an ashy white colour.[3]

From the Fiji Islands we have comparatively few butterflies ; but there are several species of Diadema of unusually pale colours, some almost white.

The Philippine Islands seem to have the peculiarity of developing metallic colours. We find there at least three species of Euplæa[4] not closely related, and all of more intense metallic lustre than their allies in other islands.

---

[1] *Ornithoptera priamus, O. helena, Papilio deiphobus, P. ulysses, P. gambrisius, P. codrus, Iphias leucippe, Euplæa prothoë, Hestia idea, Athyma jocaste, Diadema pandarus, Nymphalis pyrrhus, N. euryalus, Drusilla jairus.*

[2] "Contributions to the Theory of Natural Selection," pp. 168-173.

[3] These insects are described and figured in the " Proceedings of the Zoological Society," for 1877, p. 139. Their names are *Euplæa browni, Alcides aurora,* and *Batocera browni.*

[4] *Euplæa hewitsonii, E. diocletiana, E. lætifica.*

Here also we have one of the large yellow Ornithopteræ (*O. magellanus*), whose hind wings glow with an intense opaline lustre not found in any other species of the entire group ; and an Adolias[1] is larger and of more brilliant metallic colouring than any other species in the archipelago. In these islands also we find the extensive and wonderful genus of weevils (Pachyrhynchus), which in their brilliant metallic colouring surpass anything found in the whole eastern hemisphere, if not in the whole world.

In the Andaman Islands in the Bay of Bengal there are a considerable number of peculiar species of butter-flies differing slightly from those on the continent, and generally in the direction of paler or more conspicuous colouring. Thus two species of Papilio which on the continent have the tails black, in their Andaman re-presentatives have them either red or white-tipped.[2] Another species[3] is richly blue-banded where its allies are black ; while three species of distinct genera of Nymphalidæ[4] all differ from their allies on the continent in being of excessively pale colours as well as of some-what larger size.

In Madagascar we have the very large and singularly white-spotted *Papilio antenor ;* while species of three other genera[5] are very white or conspicuous as compared with their continental allies.

Passing to the West-Indian Islands and Central America (which latter country has formed a group of

[1] *Adolias calliphorus.*
[2] *Papilio rhodifer* (near *P. doubledayi*), and *Papilio charicles* (near *P, memnon*).   [3] *Papilio mayo.*
[4] *Euplœa andamanensis, Cethosia biblis, Cyrestis cocles.*
[5] *Danais nossima, Melanitis massoura, Diadema dexithea.*

islands in very recent times) we have similar indications. One of the largest of the Papilios inhabits Jamaica,[1] while another, the largest of its group, is found in Mexico.[2] Cuba has two of the same genus whose colours are of surpassing brilliancy;[3] while the fine genus Clothilda—confined to the Antilles and Central America — is remarkable for its rich and showy colouring.

Persons who are not acquainted with the important structural differences that distinguish these various genera of butterflies can hardly realize the importance and the significance of such facts as I have now detailed. It may be well, therefore, to illustrate them by supposing parallel cases to occur among the Mammalia. We might have, for example, in Africa, the gnus, the elands, and the buffaloes, all coloured and marked like zebras, stripe for stripe over the whole body exactly corresponding. So the hares, marmots, and squirrels of Europe might be all red with black feet, while the corresponding species of Central Asia were all yellow with black heads. In North America we might have raccoons, squirrels, and opossums, in particoloured livery of white and black, so as exactly to resemble the skunk of the same country ; while in South America they might be black with a yellow throat-patch, so as to resemble with equal closeness the tayra of the Brazilian forests. Were such resemblances to occur in anything like the number and with the wonderful accuracy of imitation met with among the Lepidoptera, they would certainly attract universal attention among naturalists, and would lead to the exhaustive

---

[1] *Papilio homerus.*   [2] *P. daunus.*   [3] *P. gundlachianus, P. villiersi.*

study of the influence of local causes in producing such startling results.

One somewhat similar case does indeed occur among the Mammalia, two singular African animals, the Aard-wolf (Proteles) and the hyæna-dog (Lycaon), both strikingly resembling hyænas in their general form as well as in their spotted markings. Belonging as they all do to the Carnivora, though to three distinct families, it seems quite an analogous case to those we have imagined ; but as the Aard-wolf and the hyæna-dog are both weak animals compared with the hyæna, the resemblance may be useful, and in that case would come under the head of mimicry. This seems the more probable because, as a rule, the colours of the Mammalia are protective, and are too little varied to allow of the influence of local causes producing any well-marked effects.

When we come to birds, however, the case is different ; for although they do not exhibit such distinct marks of the influence of locality as do butterflies— probably because the causes which determine colour are in their case more complex—yet there are distinct indications of some effect of the kind, and we must devote some little time to their consideration.

One of the most curious cases is that of the parrots of the West-Indian Islands and Central America, several of which have white heads or foreheads, occurring in two distinct genera,[1] while none of the more numerous parrots of South America are so coloured. In the small island of Dominica we have a very large and richly-

---

[1] *Pionus albifrons* and *Chrysotis senilis* (C. America), *Chrysotis sallæi* (Hayti).

coloured parrot (*Chrysotis augusta*) corresponding to the large and richly-coloured butterfly (*Papilio homerus*) of Jamaica.

The Andaman Islands are equally remarkable, at least six of the peculiar birds differing from their continental allies in being much lighter, and sometimes with a large quantity of pure white in the plumage,[1] exactly corresponding to what occurs among the butterflies.

In the Philippines this is not so marked a feature ; yet we have here the only known white-breasted king-crow (*Dicrurus mirabilis*) ; the newly discovered *Eurylæmus steerii*, wholly white beneath ; three species of Diceum, all white beneath ; several species of Parus, largely white-spotted ; while many of the pigeons have light ashy tints. The birds generally, however, have rich dark colours, similar to those which prevail among the butterflies.

In Celebes we have a swallow-shrike and a peculiar small crow allied to the jackdaw,[2] whiter than any of their allies in the surrounding islands ; but otherwise the colours of the birds call for no special remark.

In Timor and Flores we have white-headed pigeons,[3] and a long-tailed flycatcher almost entirely white.[4]

In Duke-of-York Island east of New Guinea we find that the four new species figured in the " Proceedings of the Zoological Society," for 1877, are *all* remarkable for the unusual quantity of white in their plumage. They consist of a flycatcher, a diceum, a wood-swallow, and

---

[1] *Kittacincla albiventris, Geocichla albigularis, Sturnia andamanensis, Hyloterpe grisola* var., *Ianthœnas palumboides, Osmotreron chloroptera.*

[2] *Artamus monachus, Corvus advena.*

[3] *Ptilopus cinctus, P. albocinctus.*   [4] *Tchitrea affinis,* var.

a ground pigeon ;[1] all equalling if not surpassing their nearest allies in whiteness, although some of these, from the Philippines Moluccas and Celebes, are sufficiently remarkable in this respect.

In the small Lord Howe's Island we have the recently extinct white rail (*Notornis alba*), remarkably contrasting with its allies in the larger islands of New Zealand.

We cannot, however, lay any stress on isolated examples of white colour, since these occur in most of the great continents ; but where we find a series of species of distinct genera all differing from their continental allies in a whiter coloration, as in the Andaman Islands, Duke-of-York Island, and the West Indies, and, among butterflies, in the smaller Moluccas, the Andamans, and Madagascar, we cannot avoid the conclusion that in these insular localities some general cause is at work.

There are other cases, however, in which local influences seem to favour the production or preservation of intense crimson or a very dark coloration. Thus in the Moluccas and New Guinea alone we have bright red parrots belonging to two distinct families,[2] and which therefore most probably have been independently produced or preserved by some common cause. Here, too, and in Australia we have black parrots and pigeons ;[3] and it is a most curious and suggestive fact that in another insular subregion—that of Madagascar and the Mascarene Islands—these same colours reappear in the same two groups.[4]

---

[1] *Monarcha verticalis, Diceum eximium, Artamus insignis, Phlogœnas johannœ.*

Lorius, Eos (Trichoglossidæ), *Eclectus* (Palæornithidæ).

[3] *Microglossus, Calyptorhynchus, Turacœna.*   [4] *Coracopsis, Alectrœnas.*

Sense-perception influenced by Colour of the Integuments.—Some very curious physiological facts bearing upon the presence or absence of white colours in the higher animals have lately been adduced by Dr. Ogle.[1] It has been found that a coloured or dark pigment in the olfactory region of the nostrils is essential to perfect smell, and this pigment is rarely deficient except when the whole animal is pure white. In these cases the creature is almost without smell or taste. This, Dr. Ogle believes, explains the curious case of the pigs in Virginia adduced by Mr. Darwin, white pigs being killed by a poisonous root which does not affect black pigs. Mr. Darwin imputed this to a constitutional difference accompanying the dark colour, which rendered what was poisonous to the white-coloured animals quite innocuous to the black. Dr. Ogle, however, observes that there is no proof that the black pigs eat the root, and he believes the more probable explanation to be that it is distasteful to them ; while the white pigs, being deficient in smell and taste, eat it and are killed. Analogous facts occur in several distinct families. White sheep are killed in the Tarentino by eating *Hypericum crispum*, while black sheep escape ; white rhinoceroses are said to perish from eating *Euphorbia candelabrum ;* and white horses are said to suffer from poisonous food where coloured ones escape. Now it is very improbable that a constitutional immunity from poisoning by so many distinct plants should, in the case of such widely different animals, be always correlated with the same difference of colour ; but the facts are readily understood if the senses of smell and taste are dependent on the presence

[1] "Medico-Chirurgical Transactions," vol. liii. (1870).

of a pigment which is deficient in wholly white animals. The explanation has, however, been carried a step further, by experiments showing that the absorption of odours by dead matter, such as clothing, is greatly affected by colour; black being the most powerful absorbent; then blue, red, yellow, and lastly white. We have here a physical cause for the sense-inferiority of totally white animals which may account for their rarity in nature; for few, if any, wild animals are wholly white. The head, the face, or at least the muzzle or the nose, are generally black; the ears and eyes are also often black; and there is reason to believe that dark pigment is essential to good hearing, as it certainly is to perfect vision. We can therefore understand why white cats with blue eyes are so often deaf, a peculiarity we notice more readily than their deficiency of smell or taste.

If, then, the prevalence of white coloration is generally associated with some deficiency in the acuteness of the most important senses, this colour becomes doubly dangerous; for it not only renders its possessor more conspicuous to its enemies, but at the same time makes it less ready in detecting the presence of danger. Hence, perhaps, the reason why white appears more frequently in islands, where competition is less severe and enemies less numerous and varied. Hence, also, a reason why *albinoism*, although freely occurring in captivity, never maintains itself in a wild state, while *melanism* does. The peculiarity of some islands in having all their inhabitants of dusky colours (as the Galapagos) may also perhaps be explained on the same principles; for poisonous fruits may there abound which weed out all white- or light-coloured varieties, owing to their deficiency of smell and taste. We can hardly believe, however, that this would apply to white-coloured butterflies ; and this may be a reason why the effect of an insular habitat is more marked in these insects than in birds or mammals.

It is even possible that this relation of sense-acuteness with colour may have had some influence on the development of the higher human races. If light tints of the skin were generally accompanied by some deficiency in the senses of smell, hearing, and vision, the white could never compete with the darker races so long as man was in a very low or savage condition, and wholly dependent for existence on the acuteness of his senses. But as the mental faculties became more fully developed and more important to his welfare than mere sense-acuteness, the lighter tints of skin and hair and eyes would cease to be disadvantageous whenever they were accompanied by superior brain-power. Such variations would then be preserved ; and thus may have arisen the Xanthochroic race of mankind, in which we find a high development of intellect accompanied by a slight deficiency in the acuteness of the senses as compared with the darker forms.

Relations of Insular Plants and Insects.—I have now to ask your attention to a few remarks on the peculiar relations of plants and insects as exhibited in islands.

Ever since Mr. Darwin showed the immense importance of insects in the fertilization of flowers, great attention has been paid to the subject, and the relation of these two very different classes of natural objects has been found to be more universal and more complex than could have been anticipated. Whole genera and families of plants have been so modified as, first to attract and then to be fertilized by, certain groups of insects; and this special adaptation seems in many cases to have determined the more or less wide range of the plants in question. It is also known that some species of plants can be fertilized only by particular species of insects; and the absence of these from any locality would necessarily prevent the continued existence of the plant in that area.

In this direction, I believe, will be found the clue to much of the peculiarity of the floras of oceanic islands; since the methods by which these have been stocked with plants and with insects will be often quite different. Many seeds are, no doubt, carried by oceanic currents, others probably by aquatic birds. Mr. H. N. Moseley informs me that the albatrosses, gulls, puffins, tropic birds and many others, nest inland, often amidst dense vegetation; and he believes they often carry seeds, attached to their feathers, from island to island for great distances. In the tropics they often nest on the mountains far inland, and may thus aid in the distribution even of mountain-plants. Insects, on the other hand, are mostly conveyed by aerial currents, especially by violent gales; and it may thus often happen that totally unrelated plants and insects may be brought together, in which case the former must often perish for want of suitable insects to fertilize them. This will, I think, account for the strangely fragmentary nature of these insular floras, and the great differences that often exist between those which are situated in the same

ocean ; as well as for the preponderance of certain orders
and genera.

In Mr. Pickering's valuable work on the "Geographical
Distribution of Animals and Plants " (founded on his
researches during the United States exploring expedition),
he gives a list of no less than sixty-six natural orders
of plants *unexpectedly* absent from Tahiti, or which
occur in many of the surrounding lands ; some being
abundant in other islands—as the Labiatæ at the
Sandwich Islands. In these latter islands the flora is
much richer, yet a large number of families which
abound in other parts of Polynesia are totally wanting.
Now much of the poverty and exceptional distribution
of the plants of these islands is probably due to the
great scarcity of flower-frequenting insects. Lepidoptera
and Hymenoptera are exceedingly scarce in the eastern
islands of the Pacific, and it is almost certain that many
plants which require these insects for their fertilization
have been thereby prevented from establishing themselves. In the western islands, such as the Fijis, several
species of butterflies occur in tolerable abundance, and
no doubt some flower-haunting Hymenoptera accompany
them ; and in these islands the flora appears to be much
more varied, and especially to be characterized by a
much greater variety of showy flowers, as may be seen
by examining the plates of Dr. Seeman's " Flora
Vitiensis."

Darwin and Pickering both speak of the great preponderance of ferns at Tahiti ; and Mr. Moseley, who
spent several days in the interior of the island, informs
me that "at an elevation of from 2,000 to 3,000 feet
the dense vegetation is composed almost entirely of

ferns.   A tree fern (*Alsophila tahitensis*) forms a sort of forest to the exclusion of almost every other tree, and, with huge plants of two other ferns (*Angiopteris evecta* and *Aspèlenium nidus*), forms the main mass of the vegetation." And he adds, "I have nowhere seen ferns in so great proportionate abundance." This unusual proportion of ferns is a general feature of insular as compared with continental floras ; but it has, I believe, been generally attributed to favourable conditions, especially to equable climate and perennial moisture. In this respect, however, Tahiti can hardly differ greatly from many other islands, which yet have no such vast preponderance of ferns. This is a question that cannot be decided by mere lists of species, since it is probable that in Tahiti they are less numerous than in some other islands where they form a far less conspicuous feature in the vegetation. The island most comparable with Tahiti in this respect is Juan Fernandez. Mr. Moseley writes to me :—" In a general view of any wide stretch of the densely clothed mountainous surface of the island, the ferns, both tree ferns and the unstemmed forms, are seen at once to compose a very large proportion of the mass of foliage." As to the insects of Juan Fernandez, Mr. Edwyn C. Reed, who made two visits and spent several weeks there, has kindly furnished me with some exact information. Of butterflies there is only one (*Pyrameis carie*), and that rare—a Chilian species and probably an accidental straggler. Four species of moths of moderate size were observed (all Chilian), and a few larvæ and pupæ. Of bees there were none, except one very minute species (allied to Chilicola), and of other Hymenoptera a single specimen of *Ophion luteus* a cosmopolitan ichneumon. About twenty species of flies were observed, and these formed the most prominent feature of the entomology of the island.

Now, as far as we know, this extreme entomological poverty agrees closely with that of Tahiti; and there are probably no other portions of the globe equally favoured in soil and climate, and with an equally luxuriant vegetation, where insect-life is so scantily developed. It is curious, therefore, to find that these two islands also agree in the wonderful predominance of ferns over the flowering plants—in individuals even more than in species; and there is no difficulty in connecting the two facts. The excessive minuteness and great abundance of fern-spores causes them to be far more easily distributed by winds than the seeds of flowering plants; and they are thus always ready to occupy any vacant places in suitable localities, and to compete with the less vigorous flowering plants. But where insects are so scarce, all plants which require insect-fertilization, whether constantly to enable them to produce seed at all or occasionally to keep up their constitutional vigour by crossing, must be at a great disadvantage; and thus the scanty flora which oceanic islands must always possess, peopled as they usually are by waifs and strays from other lands, is rendered still more scanty by the weeding out of all such as depend largely on insect-fertilization for their full development. It seems probable, therefore, that the preponderance of ferns in islands (considered in mass of individuals rather than in number of species) is largely due to the absence of competing phænogamous plants, and that this is in great part due to the scarcity of insects. In other oceanic islands, such as New Zealand and the Galapagos, where ferns, although tolerably abundant, form no such predominant feature in the vegetation, but where the scarcity of flower-haunting insects is almost equally marked, we find a great preponderance of small, green, or otherwise inconspicuous flowers, indicating that only such plants have been enabled to flourish there as are independent of insect-fertilization. In the Galapagos (which are perhaps even more deficient in flying insects than Juan Fernandez) this is so striking a feature that Mr. Darwin speaks of the vegetation as consisting in great part of " wretched-looking weeds," and states that " it was some time before he discovered that almost every plant was in flower at the time of his visit." He also says that he " did not see one beautiful flower " in the islands. It appears, however, that Compositæ, Leguminosæ, Rubiaceæ, and Solanaceæ form a large proportion of the flowering plants ; and as these are orders which usually require insect-fertilization, we must suppose, either that they have become modified so as to be self-fertilized, or that they are fertilized by the visits of the minute Diptera and Hymenoptera which are the only insects recorded from these islands.

In Juan Fernandez, on the other hand, there is no such total deficiency of showy flowers. I am informed by Mr. Moseley that a variety of the Magnoliaceous winter-bark abounds and has showy white flowers, and that a Bignoniaceous shrub with abundance of dark blue flowers was also plentiful ; while a white-flowered Lili-aceous plant formed large patches on the hill-sides. Besides these, there were two species of woody Com-positæ with conspicuous heads of yellow blossoms, and a species of white-flowered myrtle also abundant; so that, on the whole, flowers formed a rather conspicuous feature in the aspect of the vegetation of Juan Fernandez.

But this fact—which at first sight seems entirely at variance with the view we are upholding of the important relation between the distribution of insects and plants—is well explained by the existence of two species of humming-birds in Juan Fernandez, which, in their visits to these large and showy flowers, fertilize them as effectually as bees, moths, or butterflies. Mr. Moseley informs me that "these humming-birds are *extraordinarily abundant,* every tree or bush having one or two darting about it." He also observed that "nearly all the specimens killed had the feathers round the base of the bill and front of the head clogged and coloured yellow with pollen." Here, then, we have the clue to the perpetuation of large and showy flowers in Juan Fernandez; while the total absence of humming-birds in the Galapagos may explain why no such large-flowered plants have been able to establish themselves in those equatorial islands.

This leads to the observation that many other groups of birds also, no doubt, aid in the fertilization of flowers. I have often observed the beaks and faces of the brush-tongued lories of the Moluccas covered with pollen; and Mr. Moseley noted the same fact in a species of Artamus, or swallow-shrike, shot at Cape York, showing that this genus also frequents flowers and aids in their fertilization. In the Australian region we have the immense group of the Meliphagidæ, which all frequent flowers; and as these range over all the islands of the Pacific, their presence will account for a certain proportion of showy flowers being found there, such as the scarlet Metrosideros, one of the few conspicuous flowers in Tahiti. In the Sandwich Islands, too, there are forests of Metrosideros ; and Mr. Charles Pickering writes me, that they are visited by honey-sucking birds, one of which is captured by sweetened bird-lime, against which it thrusts its extensile tongue. I am also informed that a considerable number of flowers are occasionally fertilized by humming-birds in North America ; so that there can, I think, be little doubt that birds play a much more important part in this respect than has hitherto been imagined. It is not improbable that in Tropical America, where the humming-bird family is so enormously developed, many flowers will be found to be expressly adapted to fertilization by them, just as so many in our own country are specially adapted to the visits of certain families or genera of insects.[1]

It must also be remembered, as Mr. Moseley has suggested to me, that a flower which has acquired a brilliant colour to attract insects might, on transference to another country and becoming so modified as to be capable of self-fertilization, retain the coloured petals for

[1] The probable influence of fertilization by birds on the flowers of the Auckland Isles has been referred to at p. 238. Mr. Darwin, in his book on *Cross and Self-Fertilisation of Plants* (p. 371), gives in a note numerous cases in which birds are known to fertilise flowers, the most important being that of several species of *Abutilon* in South Brazil, which, according to Fritz Müller, are sterile unless fertilised by humming-birds. This proves, not only that birds fertilise flowers in the same manner as insects, but that the two classes of organisms have become so correlated as to be mutually necessary to each other ; and it completely justifies us in imputing the fertilization of flowers to flower-frequenting birds wherever these are present and suitable insects are notoriously scarce, as is the case in so many of the islands here referred to.

an indefinite period. Such is probably the explanation
of the Pelargonium of Tristan d'Acunha, which forms
masses of bright colour near the shore during the
flowering season ; while most of the other plants of the
island have colourless flowers in accordance with the
almost total absence of winged insects. The presence
of many large and showy flowers among the indigenous
flora of St. Helena must be an example of a similar
persistence. Mr. Melliss indeed states it to be " a
remarkable peculiarity that the indigenous flowers are,
with very slight exceptions, all perfectly colourless ; " [1]
but although this may apply to the general aspect of
the remains of the indigenous flora, it is evidently not
the case as regards the *species*, since the interesting
plates of Mr. Melliss's volume show that about one
third of the indigenous flowering plants have more or
less coloured or conspicuous flowers, while several of
them are exceedingly showy and beautiful. Among
these are a Lobelia, three Wahlenbergias, several Compositæ, and especially the handsome red flowers of the
now almost extinct forest-trees, the ebony and redwood
(species of Melhania, Byttneriaceæ). We have every
reason to believe, however, that when St. Helena was
covered with luxuriant forests, and especially at that
remote period when it was much more extensive than
it is now, it must have supported a certain number of
indigenous birds and insects, which would have aided in
the fertilization of these gaily-coloured flowers. The
researches of Dr. Hermann Müller have shown us by
what minute modifications of structure or of function,
many flowers are adapted for partial insect and self-

[1] Melliss's *St. Helena* p. 226, note.

fertilization in various degrees; so that we have no difficulty in understanding how, as the insects diminished and finally disappeared, self-fertilization may have become the rule, while the large and showy corollas remain to tell us plainly of a once different state of things. ject is the presence of arborescent forms of Compositæ in so many of the remotest oceanic islands. They occur in the Galapagos, in Juan Fernandez, in St. Helena, in the Sandwich Islands, and in New Zealand ; but they are not directly related to each other ; representatives of totally different tribes of this extensive order becoming arborescent in each group of islands. The immense range and almost universal distribution of the Compositæ is due to the combination of a great facility of distribution (by their seeds) with a great attractiveness to insects ; and to the capacity of being fertilized by a variety of species of all orders, and especially by flies and small beetles. Thus they would be among the earliest of flowering plants to establish themselves on oceanic islands ; but where insects of all kinds were very scarce, it would be an advantage to gain increased size and longevity, so that fertilization at an interval of several years might suffice for the continuance of the species. The arborescent form would combine with increased longevity the advantage of increased size in the struggle for existence with ferns and other early colonists ; and these advantages have led to its being independently produced in so many distant localities, whose chief feature in common is their remoteness from continents and the extreme poverty of their insect life.

As the sweet odours of flowers are known to act in

combination with their colours, as an attraction to insects,
it might be anticipated that where colour was deficient
scent would be so also. On applying to my friend Sir
Joseph Hooker for information as to the odoriferous
qualities of New-Zealand plants, he informed me, that
the New-Zealand flora is, speaking generally, as strikingly deficient in sweet odours as it is in conspicuous
colours. Whether this peculiarity occurs in other islands
I have not been able to obtain information ; but we may
certainly expect to find it where colour is so strikingly
deficient as in the flora of the Galapagos Islands.

Another question which here comes before us, is the
origin and meaning of the odoriferous glands of leaves.
Sir Joseph Hooker informed me that not only are New-
Zealand plants deficient in bright coloured and sweet-
smelling flowers, but equally so in scented leaves. This
led me to think that perhaps such leaves were in some
way an additional attraction to insects—though it is not
easy to understand how this could be, except by adding
a general attraction to the special attraction of the
flowers, or by supporting the larvæ which, as perfect insects, aid in fertilization. Mr. Darwin, however, informs
me that he considers that leaf-glands bearing essential oils
are a protection against the attacks of insects where these
abound, and would thus not be required in countries
where insects were very scarce. But it seems opposed
to this view that highly aromatic plants are characteristic of deserts all over the world, and in such places
insects are not abundant. Mr. Stainton informs me that
the aromatic Labiatæ enjoy no immunity from insect
attacks. The bitter leaves of the cherry-laurel are often
eaten by the larvæ of moths that abound on our fruit-

trees ; while in the Tropics the leaves of the orange tribe are favourites with a large number of lepidopterous larvæ; and our northern firs and pines, although abounding in a highly aromatic resin, are very subject to the attacks of beetles. My friend Dr. Richard Spruce—who while travelling in South America allowed nothing connected with plant-life to escape his observation—informs me that trees whose leaves have aromatic and often resinous secretions in immersed glands abound in the plains of tropical America, and that such are in great part, if not wholly, free from the attacks of leaf-eating ants, except where the secretion is only slightly bitter, as in the orange tribe, orange-trees being sometimes entirely denuded of their leaves in a single night. Aromatic plants abound in the Andes up to about 13,000 feet, as well as in the plains, but hardly more so than in Central and Southern Europe. They are perhaps more plentiful in the dry mountainous parts of Southern Europe ; and as neither here nor in the Andes do leaf-eating ants exist, Dr. Spruce infers that, although in the hot American forests where such ants swarm the oil-bearing glands serve as a protection, yet they were not originally acquired for that purpose. Near the limits of perpetual snow on the Andes such plants as occur are not, so far as Dr. Spruce has observed, aromatic ; and as plants in such situations can hardly depend on insect visits for their fertilization, the fact is comparable with that of the flora of New Zealand, and would seem to imply some relation between the two phenomena, though what it exactly is cannot yet be determined.

I trust I have now been able to show you that there

are a number of curious problems lying as it were on the outskirts of biological inquiry which well merit attention, and which may lead to valuable results. But these problems are, as you see, for the most part connected with questions of locality, and require full and accurate knowledge of the productions of a number of small islands and other limited areas, and the means of comparing them one with the other. To make such comparisons, however, is now quite impossible. No museum contains any fair representation of the productions of these localities; and such specimens as do exist, being scattered through the general collection, are almost useless for this special purpose. If, then, we are to make any progress in this inquiry it is absolutely essential that some collectors should begin to arrange their cabinets primarily on a geographical basis, keeping together the productions of every island or group of islands, and of such divisions of each continent as are found to possess any special or characteristic fauna or flora. We shall then be sure to detect many unsuspected relations between the animals and plants of certain localities, and we shall become much better acquainted with those complex reactions between the vegetable and animal kingdoms, and between the organic world and the inorganic, which have almost certainly played an important part in determining many of the most conspicuous features of living things.

RISE AND PROGRESS OF MODERN VIEWS AS TO THE
ANTIQUITY AND ORIGIN OF MAN.

I now come to a branch of our subject which I would
gladly have avoided touching on ; but as the higher
powers of the British Association have decreed that I
should preside over the Anthropological Department, it
seems proper that I should devote some portion of my
address to matters more immediately connected with the
special study to which that Department is devoted.

As my own knowledge of and interest in Anthropology
is confined to the great outlines rather than to the special
details of the science, I propose to give a very brief and
general sketch of the modern doctrine as to the Antiquity
and Origin of Man, and to suggest certain points of difficulty which have not, I think, yet received sufficient
attention.

Many now living remember the time (for it is little
more than twenty years ago) when the antiquity of man,
as now understood, was universally discredited. Not
only theologians, but even geologists then taught us, that
man belonged altogether to the existing state of things ;
that the extinct animals of the Tertiary period had finally
disappeared, and that the earth's surface had assumed
its present condition before the human race first came
into existence. So prepossessed were even scientific men
with this idea—which yet rested on purely negative
evidence, and could not be supported by any arguments
of scientific value—that numerous facts which had been
presented at intervals for half a century, all tending to
prove the existence of man at very remote epochs, were

silently ignored; and, more than this, the detailed statements of three distinct and careful observers confirming each other, were rejected by a great scientific Society as too improbable for publication, only because they proved (if they were true) the coexistence of man with extinct animals. [1]

But this state of belief in opposition to facts, could not long continue. In 1859 a few of. our most eminent geologists examined for themselves into the alleged occurrence of flint implements in the gravels of the north of France, which had been made public fourteen years before, and found them strictly correct. The caverns of Devonshire were about the same time carefully examined by equally eminent observers, and were found fully to bear out the statements of those who had published their results eighteen years before. Flint implements began to be found in all suitable localities in the south of England, when carefully searched for, often in gravels of equal antiquity with those of France. Caverns giving evidence of human occupation at various remote periods were explored in Belgium and the south of France—lake-dwellings were examined in Switzerland —refuse-heaps in Denmark—and thus a whole series of remains have been discovered carrying back the history of mankind from the earliest historic periods to a long distant past.

The antiquity of the races thus discovered cannot be measured in years; but it may be approximately deter-

---

[1] In 1854 (?) a communication from the Torquay Natural-History Society confirming previous accounts by Mr. Godwin-Austen, Mr. Vivian, and the Rev. Mr. M'Enery, that worked flints occurred in Kent's Hole with remains of extinct species, was rejected as too improbable for publication.

mined by the successively earlier and earlier stages of civilization through which we can trace them, and by the changes in physical geography and of animal and vegetable life that have since occurred. As we go back metals soon disappear, and we find only tools and weapons of stone and of bone. The stone weapons get ruder and ruder; pottery, and then the bone implements, cease to occur; and in the earliest stage we find only chipped flints of rude design, though still of unmistakably human workmanship. In like manner domestic animals disappear as we go backward; and though the dog seems to have been the earliest, it is doubtful whether the makers of the ruder flint implements of the gravels possessed even this. Still more important as a measure of time are the changes in the distribution of animals, indicating changes of climate, which have occurred during the human period. At a comparatively recent epoch in the record of prehistoric times we find that the Baltic was far salter than it is now and produced abundance of oysters, and that Denmark was covered with pine forests inhabited by Capercailzies, such as now only occur further north in Norway. A little earlier we find that reindeer were common even in the south of France; and still earlier this animal was accompanied by the mammoth and woolly rhinoceros, by the arctic glutton, and by huge bears and lions of extinct species. The presence of such animals implies a change of climate; and both in the caves and gravels we find proofs of a much colder climate than now prevails in Western Europe. Even more remarkable are the changes of the earth's surface which have been effected during man's occupation of it. Many extensive valleys in England and France are believed by the best observers to have been deepened at least a hundred feet ; caverns now far out of the reach of any stream must for a long succession of years have had streams flowing through them, at least in times of floods ; and this often implies that vast masses of solid rock have since been worn away. In Sardinia land has risen at least 300 feet since men lived there who made pottery and probably used fishing-nets ;[1] while in Kent's Cavern remains of man are found buried beneath two separate beds of stalagmite, each having a distinct texture, and each covering a deposit of cave-earth having well-marked differential characters, while each contains a distinct assemblage of extinct animals.

Such, briefly, are the results of the evidence that has been rapidly accumulating for about fifteen years, as to the antiquity of man ; and it has been confirmed by so many discoveries of a like nature in all parts of the globe, and especially by the comparison of the tools and weapons of prehistoric man with those of modern savages (so that the use of even the rudest flint implements has become quite intelligible), that we can hardly wonder at the vast revolution effected in public opinion. Not only is the belief in man's vast and still unknown antiquity universal among men of science, but it is hardly disputed by any well-informed theologian ; and the present generation of science-students must, we should think, be somewhat puzzled to understand what there was in the earliest discoveries that should have aroused such general opposition, and been met with such universal incredulity.

But the question of the mere " Antiquity of Man "

[1] Lyell's *Antiquity of Man*, fourth edition, p. 115.

almost sank into insignificance at a very early period of the inquiry, in comparison with the far more momentous and more exciting problem of the development of man from some lower animal form, which the theories of Mr. Darwin and of Mr. Herbert Spencer soon showed to be inseparably bound up with it. This has been, and to some extent still is, the subject of fierce conflict; but the controversy as to the fact of such development is now almost at an end, since one of the most talented representatives of Catholic theology, and an anatomist of high standing—Professor Mivart—fully adopts it as regards physical structure, reserving his opposition for those parts of the theory which would deduce man's whole intellectual and moral nature from the same source and by a similar mode of development.

Never, perhaps, in the whole history of science or philosophy has so great a revolution in thought and opinion been effected as in the twelve years from 1859 to 1871, the respective dates of publication of Mr. Darwin's *Origin of Species* and *Descent of Man*. Up to the commencement of this period the belief in the independent creation or origin of the species of animals and plants, and the very recent appearance of man upon the earth, were, practically, universal. Long before the end of it these two beliefs had utterly disappeared, not only in the scientific world, but almost equally so among the literary and educated classes generally. The belief in the independent origin of man held its ground somewhat longer; but the publication of Mr. Darwin's great work gave even that its death-blow, for hardly any one capable of judging of the evidence now doubts the derivative nature of man's bodily structure as a whole, although many believe that his mind, and even some of his physical characteristics, may be due to the action of other forces than have acted in the case of the lower animals.

We need hardly be surprised, under these circumstances, if there has been a tendency among men of science to pass from one extreme to the other ; from a profession (so few years ago) of total ignorance as to the mode of origin of all living things, to a claim to almost complete knowledge of the whole progress of the universe, from the first speck of living protoplasm up to the highest development of the human intellect. Yet this is really what we have seen in the last sixteen years. Formerly difficulties were exaggerated, and it was asserted that we had not sufficient knowledge to venture on any generalizations on the subject. Now difficulties are set aside, and it is held that our theories are so well established and so far-reaching, that they explain and comprehend all nature. It is not long ago (as I have already reminded you) since *facts* were contemptuously ignored, because they favoured our now popular views ; at the present day it seems to me that facts which oppose them hardly receive due consideration. And as opposition is the best incentive to progress, and it is not well even for the best theories to have it all their own way, I propose to direct your attention to a few such facts, and to the conclusions that seem fairly deducible from them.

Indications of Man's Extreme Antiquity.—It is a curious circumstance that, notwithstanding the attention that has been directed to the subject in every part of the world, and the numerous excavations connected with railways and mines which have offered such facilities for geological discovery, no advance whatever has been made for a considerable number of years in detecting the time or mode of man's origin. The Palæolithic flint weapons first discovered in the North of France more than thirty years ago, are still the oldest undisputed proofs of man's existence ; and amid the countless relics of a former world that have been brought to light, no evidence of any one of the links that must have connected man with the lower animals has yet appeared.

It is, indeed, well known that negative evidence in geology is of very slender value ; and this is, no doubt, generally the case. The circumstances here are, however, peculiar, for many converging lines of evidence show that, on the theory of development by the same laws which have determined the development of the lower animals, man must be immensely older than any traces of him yet discovered. As this is a point of great interest we must devote a few moments to its consideration.

1. The most important difference between man and such of the lower animals as most nearly approach him is undoubtedly in the bulk and development of his brain, as indicated by the form and capacity of the cranium. We should therefore anticipate that these earliest races, who were contemporary with the extinct animals and used rude stone weapons, would show a marked deficiency in this respect. Yet the oldest known crania (those of the Engis and Cro-Magnon caves) show no marks of degradation. The former does not present so low a type as that of most existing savages, but is (to use the words of Prof. Huxley) "a fair average human skull, which might have belonged to a philosopher, or

might have contained the thoughtless brains of a savage."
The latter are still more remarkable, being unusually
large and well formed. Dr. Pruner-Bey states that they
surpass the average of modern European skulls in
capacity, while their symmetrical form without any trace
of prognathism, compares favourably not only with those
of the foremost savage races, but with many civilised
nations of modern times.

One or two other crania of much lower type, but of less
antiquity than this, have been discovered ; but they in
no way invalidate the conclusion which so highly developed a form at so early a period implies, viz., that we
have as yet made a hardly perceptible step towards
the discovery of any earlier stage in the development
of man.

2. This conclusion is supported and enforced by the
nature of many of the works of art found even in the
oldest cave-dwellings. The flints are of the old chipped
type, but they are formed into a large variety of tools
and weapons—such as scrapers, awls, hammers, saws,
lances, &c., implying a variety of purposes for which
these were used, and a corresponding degree of mental
activity and civilization. Numerous articles of bone
have also been found, including well-formed needles ;
implying that skins were sewn together, and perhaps
even textile materials woven into cloth. Still more
important are the numerous carvings and drawings representing a variety of animals, including horses, reindeer, and even a mammoth, executed with considerable
skill on bone, reindeer-horns, and mammoth-tusks.
These, taken together, indicate a state of civilization
much higher than that of the lowest of our modern

savages, while they are quite compatible with a considerable degree of mental advancement, and lead us to believe that the crania of Engis and Cro-Magnon are not exceptional, but fairly represent the characters of the race. If we further remember that these people lived in Europe under the unfavourable conditions of a sub-Arctic climate, we shall be inclined to agree with Dr. Daniel Wilson, that it is far easier to produce evidences of deterioration than of progress, in instituting a comparison between the contemporaries of the mammoth and later prehistoric races of Europe or savage nations of modern times.[1]

3. Yet another important line of evidence as to the extreme antiquity of the human type has been brought prominently forward by Prof. Mivart.[2] He shows, by a careful comparison of all parts of the structure of the body, that man is related not to any one, but almost equally to many of the existing apes—to the orang, the chimpanzee, the gorilla, and even to the gibbons—in a variety of ways ; and these relations and differences are so numerous and so diverse that, on the theory of evolution, the ancestral form which ultimately developed into man must have diverged from the common stock whence all these various forms and their extinct allies originated. But so far back as the Miocene deposits of Europe we find the remains of apes allied to these various forms, and especially to the gibbons ; so that in all probability the special line of variation which led up to man branched off at a still earlier period.   And these early forms, being the initiation of a far higher type, and

[1] *Prehistoric Man*, 3rd edit. vol. i. p. 117.
Man and Apes, pp. 171-193.

having to develop by natural selection into so specialized and altogether distinct a creature as man, must have risen at a very early period into the position of a dominant race, and spread in dense waves of population over all suitable portions of the great continent— for this, on Mr. Darwin's hypothesis, is essential to developmental progress through the agency of natural selection.

Under these circumstances we might certainly expect to find some relics of these earlier forms of man along with those of animals, which were presumably less abundant. Negative evidence of this kind is not very weighty, but still it has *some* value. It has been suggested that as apes are mostly tropical, and anthropoid apes are now confined almost exclusively to the vicinity of the equator, we should expect the ancestral forms of man to have inhabited these same localities—West Africa and the Malay Islands. But this objection is hardly valid, because existing anthropoid apes are wholly dependent on a perennial supply of easily accessible fruits, which is only found near the equator ; while not only had the south of Europe an almost tropical climate in Miocene times, but we must suppose even the earliest ancestors of man to have been terrestrial and omnivorous, since it must have taken ages of slow modification to have produced the perfectly erect form, the short arms, and the wholly non-prehensile foot,[1] which so strongly differentiate man from the arboreal apes.

[1] The common statement of travellers as to savages having great prehensile power in the toes, has been adopted by some naturalists as indicating an approach to the apes. But this notion is founded on a complete misconception. Savages pick up objects with their feet, it is true, but always by a lateral motion of the toes, which we should equally possess if we never wore shoes or

The conclusion which I think we must arrive at is, that if man has been developed from a common ancestor with all existing apes, *and by no other agencies than such as have affected their development*, then he must have existed, in something approaching his present form, during the tertiary period—and not merely existed, but predominated in numbers, wherever suitable conditions prevailed. If then, continued researches in all parts of Europe and Asia fail to bring to light any proofs of his presence, it will be at least a presumption that he came into existence at a much later date, and by a much more rapid process of development. In that case it will be a fair argument that, just as he is in his mental and moral nature, his capacities and aspirations, so infinitely raised above the brutes, so his origin is due, in part, to distinct and higher agencies than such as have affected their development.

Antiquity of Intellectual Man.—There is yet another line of inquiry bearing upon this subject to which I wish to call your attention. It is a somewhat curious fact that, while all modern writers admit the great antiquity of man, most of them maintain the very recent development of his intellect, and will hardly contemplate the possibility of men equal in mental capacity to ourselves having existed in prehistoric times. This question is generally assumed to be settled by such relics as have been preserved of the manufactures of the older races, showing a lower and lower state of the arts; by the successive disappearance in early times of iron,

stockings. In no savage have I ever seen the slightest approach to opposability of the great toe, which is the essential distinguishing feature of apes; nor have I ever seen it stated that any variation in this direction has been detected in the anatomical structure of the foot of the lower races.

bronze, and pottery; and by the ruder forms of the older flint implements. The weakness of this argument has been well shown by Mr. Albert Mott in his very original but little known presidential address to the Literary and Philosophical Society of Liverpool in 1873. He maintains that "our most distant glimpses of the past are still of a world peopled as now with men both civilised and savage," and "that we have often entirely misread the past by supposing that the outward signs of civilisation must always be the same, and must be such as are found among ourselves." In support of this view he adduces a variety of striking facts and ingenious arguments, a few of which I will briefly summarize.

Sculptures on Easter Island.—On one of the most remote islands of the Pacific—Easter Island, 2,000 miles from South America, 2,000 from the Marquesas, and more than 1,000 from the Gambier Islands, are found hundreds of gigantic stone images, now mostly in ruins. They are often forty feet high, while some seem to have been much larger, the crowns on their heads, cut out of a red stone, being sometimes ten feet in diameter, while even the head and neck of one is said to have been twenty feet high.[1] These images once all stood erect on extensive stone platforms.

The island containing these remarkable works of art has only an area of about thirty square miles, or considerably less than Jersey. Now as one of the smallest images (eight feet high) weighs four tons, the largest must weigh over a hundred tons, if not much more; and the existence of such vast works implies a large population, abundance of food, and an established government. Yet

[1] *Journ. of Roy. Geog. Soc.* 1870, pp. 177, 178.

how could these coexist on a mere speck of land wholly cut off from the rest of the world ? Mr. Mott maintains that these facts necessarily imply the power of regular communication with larger islands or a continent, the arts of navigation, and a civilisation much higher than now exists in any part of the Pacific. Very similar remains in other islands scattered widely over the Pacific add weight to this argument.

North American Earthworks.—The next example is that of the ancient mounds and earthworks of the North American continent, the bearing of which is even more significant. Over the greater part of the extensive Mississippi valley, four well-marked classes of these earthworks occur. Some are camps, or works of defence, situated on bluffs, promontories, or isolated hills ; others are vast inclosures in the plains and lowlands, often of geometric forms, and having attached to them roadways or avenues often miles in length ; a third are mounds corresponding to our tumuli, often seventy to ninety feet high, and some of them covering acres of ground ; while a fourth group consists of representations of various animals modelled in relief on a gigantic scale, and occurring chiefly in an area somewhat to the northwest of the other classes, in the plains of Wisconsin.

The first class—the camps or fortified inclosures—resemble in general features the ancient camps of our own islands, but far surpass them in extent. Fort Hill, in Ohio, is surrounded by a wall and ditch a mile and a half in length, part of the way cut through solid rock. Artificial reservoirs for water were made within it, while at one extremity, on a more elevated point, a keep is constructed with its separate defences and water-

reservoirs. Another, called Clark's Work, in the
Scioto valley, which seems to have been a fortified
town, incloses an area of 127 acres, the embankments
measuring three miles in length, and containing not
less than three million cubic feet of earth. This area
incloses numerous sacrificial mounds and symmetrical
earthworks, in which many interesting relics and works
of art have been found.

The second class—the sacred inclosures, may be
compared for extent and arrangement with Avebury
or Carnak, but are in some respects even more remarkable. One of these at Newark, Ohio, covers an
area of several miles, with its connected groups of circles,
octagons, squares, ellipses, and avenues on a grand scale,
and formed by embankments from twenty to thirty feet
in height. Other similar works occur in different parts
of Ohio ; and by accurate survey it is found, not only
that the circles are true, though some of them are one-
third of a mile in diameter, but that other figures are
truly square, each side being over 1,000 feet long ; and,
what is still more important, the dimensions of some of
these geometrical figures in different parts of the
country and seventy miles apart, are identical. Now
this proves the use, by the builders of these works, of
some standard measures of length ; while the accuracy
of the squares, circles, and, in a less degree, of the
octagonal figures, shows a considerable knowledge of
rudimentary geometry and some means of measuring
angles. The difficulty of drawing such figures on a
large scale is much greater than any one would imagine
who has not tried it ; and the accuracy of these is far
beyond what is necessary to satisfy the eye. We must

therefore impute to the builders the wish to make these figures as accurate as possible ; and this wish is a greater proof of habitual skill and intellectual advancement than even the ability to draw such figures. If, then, we take into account this ability and this love of geometric truth, and further consider the dense population and civil organisation implied by the construction of such extensive systematic works, we must allow that these ancient people had reached the earlier stages of a civilisation of which no traces existed among the savage tribes who alone occupied the country when first visited by Europeans.

The animal mounds are of comparatively less importance for our present purpose, as they imply a somewhat lower grade of advancement ; but the sepulchral and sacrificial mounds exist in vast numbers, and their partial exploration has yielded a quantity of articles and works of art which throw some further light on the peculiarities of this mysterious people. Most of these mounds contain a large concave hearth or basin of burnt clay, of perfectly symmetrical form, on which are found deposited more or less abundant relics, all bearing traces of the action of fire. We are therefore only acquainted with such articles as are practically fire-proof, or have accidentally escaped combustion. These consist of bone and copper implements and ornaments, disks, and tubes ; pearl, shell, and silver beads, more or less injured by the fire ; ornaments cut in mica ; ornamental pottery ; and numbers of elaborate carvings in stone, mostly forming pipes for smoking.[1] The metallic articles are

[1] Woven cloth, apparently of flax or hemp, as well as gauges supposed to have been used to regulate the thickness of the thread, have also been found

all formed by hammering, but the execution is very
good; plates of mica are found cut into scrolls and
circles; the pottery, of which very few remains have
been found, is far superior to that of any of the Indian
tribes, since Dr. Wilson is of opinion that it must have
been formed on a wheel, as it is often of uniform thickness throughout (sometimes not more than one-sixth of
an inch), polished, and ornamented with scrolls and
figures of birds and flowers in delicate relief. But the
most instructive objects are the sculptured stone pipes,
representing not only various easily recognizable animals, but also human heads, so well executed that they
appear to be portraits. Among the animals, not only
are such native forms as the panther, bear, otter, wolf,
beaver, raccoon, heron, crow, turtle, frog, rattlesnake,
and many others well represented, but also the manatee,
which perhaps then ascended the Mississippi as it now
does the Amazon, and the toucan, which could hardly
have been obtained nearer than Mexico. The sculptured
heads are especially remarkable, because they present to
us the features of an intellectual and civilised people.
The nose in some is perfectly straight, and neither
prominent nor dilated; the mouth is small, and the
lips thin; the chin and upper lip are short, contrasting
with the ponderous jaw of the modern Indian, while
the cheek-bones present no marked prominence. Other
examples have the nose somewhat projecting at the apex
in a manner quite unlike the features of any American
indigenes; and although there are some which show a
much coarser face, it is very difficult to see in any of

in several of the mounds of Ohio.   (Foster's *Prehistoric Races of the United
States*, 1873, pp. 225-229.)

them that close resemblance to the Indian type which these sculptures have been said to exhibit. The few authentic crania from the mounds present corresponding features, being far more symmetrical and better developed in the frontal region than those of any American tribes, although somewhat resembling them in the occipital outline ;[1] while one was described by its discoverer (Mr. W. Marshall Anderson) as a "beautiful skull, worthy of a Greek."

The antiquity of this remarkable race may perhaps not be very great as compared with the prehistoric man of Europe, although the opinion of some writers on the subject seems affected by that "parsimony of time" on which the late Sir Charles Lyell so often dilated. The mounds are all overgrown with dense forest, and one of the large trees was estimated to be 800 years old, while other observers consider the forest growth to indicate an age of at least 1,000 years. But it is well known that it requires several generations of trees to pass away before the growth on a deserted clearing comes to correspond with that of the surrounding virgin forest, while this forest, once established, may go on growing for an unknown number of thousands of years. The 800 or 1,000 years estimate from the growth of existing vegetation is a minimum which has no bearing whatever on the actual age of these mounds ; and we might almost as well attempt to determine the time of the glacial epoch from the age of the pines or oaks which now grow on the moraines.

The important thing for us, however, is that when North America was first settled by Europeans, the Indian

[1] Wilson's *Prehistoric Man*, 3rd edit. vol. ii. pp. 123-130.

tribes inhabiting it had no knowledge or tradition of any preceding race of higher civilisation than themselves. Yet we find that such a race existed; that they must have been populous and have lived under some established government; while there are signs that they practised agriculture largely, as, indeed, they must have done to have supported a population capable of executing such gigantic works in such vast profusion; for it is stated that the mounds and earthworks of various kinds in the state of Ohio alone, amount to between eleven and twelve thousand. In their habits, customs, religion, and arts, they differed strikingly from all the Indian tribes; while their love of art and of geometric forms, and their capacity for executing the latter upon so gigantic a scale, render it probable that they were a really civilised people, although the form their civilisation took may have been very different from that of later peoples, subject to very different influences and the inheritors of a longer series of ancestral civilisations. We have here, at all events, a striking example of the transition, over an extensive country, from comparative civilisation to comparative barbarism, the former leaving no tradition and hardly any trace of its influence on the latter.

As Mr. Mott well remarks :—Nothing can be more striking than the fact that Easter Island and North America both give the same testimony as to the origin of the savage life found in them, although in all circumstances and surroundings the two cases are so different. If no stone monuments had been constructed in Easter Island, or mounds containing a few relics saved from fire, in the United States, we might never have suspected the existence of these ancient peoples. He argues, therefore, that it is very easy for the records of an ancient nation's life entirely to perish or to be hidden from observation. Even the arts of Nineveh and Babylon were unknown only a generation ago, and we have only just discovered the facts about the mound-builders of North America.

But other parts of the American continent exhibit parallel phenomena. Recent investigations show that in Mexico, Central America, and Peru the existing race of Indians has been preceded by a distinct and more civilised race. This is proved by the sculptures of the ruined cities of Central America, by the more ancient terra-cottas and paintings of Mexico, and by the oldest portrait-pottery of Peru. All alike show markedly non-Indian features, while they often closely resemble modern European types. Ancient crania, too, have been found in all these countries, presenting very different characters from those of any of the existing indigenous races of America.[1]

The Great Pyramid.—There is one other striking example of a higher being succeeded by a lower degree of knowledge, which is in danger of being forgotten because it has been made the foundation of theories which seem wild and fantastic, and are probably in great part erroneous. I allude to the Great Pyramid of Egypt, whose form, dimensions, structure, and uses have recently been the subject of elaborate works by Prof. Piazzi Smyth. Now the admitted facts about the pyramid are so interesting and so apposite to the subject

---

[1] Wilson's *Prehistoric Man*, 3rd edit. vol. ii. pp. 125, 144.

we are considering, that I beg to recall them to your attention. Most of you are aware that this pyramid has been carefully explored and measured by successive Egyptologists, and that the dimensions have lately become capable of more accurate determination, owing to the discovery of some of the original casing-stones, and the clearing away of the earth from the corners of the foundation showing the sockets in which the corner-stones fitted. Prof. Smyth devoted many months of work with the best instruments, in order to fix the dimensions and angles of all accessible parts of the structure ; and he has carefully determined these by a comparison of his own and all previous measures, the best of which agree pretty closely with each other. The results arrived at are :—

1. That the pyramid is truly square, the sides being equal and the angles right angles.

2. That the four sockets on which the four first stones of the corners rested, are truly on the same level.

3. That the directions of the sides are accurately to the four cardinal points.

4. That the vertical height of the pyramid bears the same proportion to its circumference at the base, as the radius of a circle does to its circumference.

Now all these measures, angles, and levels are accurate, not as an ordinary surveyor or builder could make them, but to such a degree as requires the very best modern instruments and all the refinements of geodetical science to discover any error at all. In addition to this we have the wonderful perfection of the workmanship in the interior of the pyramid, the passages and chambers being lined with huge blocks of stones fitted with the utmost accuracy, while every part of the building exhibits the highest structural science.

In all these respects this largest pyramid surpasses every other in Egypt. Yet it is universally admitted to be the oldest, and also the oldest historical building in the world.

Now these admitted facts about the Great Pyramid are surely remarkable, and worthy of the deepest consideration. They are facts which, in the pregnant words of the late Sir John Herschel, "according to received theories ought not to happen," and which, he tells us, should therefore be kept ever present to our minds, since "they belong to the class of facts which serve as the clue to new discoveries." According to modern theories, the higher civilisation is ever a growth and an outcome from a preceding lower state ; and it is inferred that this progress is visible to us throughout all history and in all material records of human intellect. But here we have a building which marks the very dawn of history, which is the oldest authentic monument of man's genius and skill, and which, instead of being far inferior, is very much superior to all which followed it. Great men are the products of their age and country, and the designer and constructors of this wonderful monument could never have arisen among an unintellectual and half-barbarous people. So perfect a work implies many preceding less perfect works which have disappeared. It marks the culminating point of an ancient civilisation, of the early stages of which we have no trace or record whatever.

The three cases to which I have now adverted (and there are many others) seem to require for their satisfactory interpretation a somewhat different view of human progress from that which is now generally accepted. Taken in connection with the great intellectual power of the ancient Greeks—which Mr. Galton believes to have been far above that of the average of any modern nation—and the elevation, at once intellectual and moral, displayed in the writings of Confucius, Zoroaster, and the Vedas, they point to the conclusion that, while in material progress there has been a tolerably steady advance, man's intellectual and moral development reached almost its highest level in a very remote past. The lower, the more animal, but often the more energetic types have, however, always been far the more numerous ; hence such established societies as have here and there arisen under the guidance of higher minds have always been liable to be swept away by the incursions of barbarians. Thus in almost every part of the globe there may have been a long succession of partial civilisations, each in turn succeeded by a period of barbarism ; and this view seems supported by the occurrence of degraded types of skull along with such "as might have belonged to a philosopher," at a time when the mammoth and the reindeer inhabited southern France.

Nor need we fear that there is not time enough for the rise and decay of so many successive civilisations as this view would imply ; for the opinion is now gaining ground among geologists that palæolithic man was really preglacial, and that the great gap (marked alike by a change of physical conditions and of animal life) which in Europe always separates him from his neolithic successor, was caused by the coming on and passing away of the great ice age.

If the views now advanced are correct, many, perhaps most, of our existing savages are the successors of higher races ; and their arts, often showing a wonderful similarity in distant continents, may have been derived from a common source among more civilised peoples.

Conclusion.—I must now conclude this very imperfect sketch of a few of the offshoots from the great tree of Biological study. It will, perhaps, be thought by some that my remarks have tended to the depreciation of our science, by hinting at imperfections in our knowledge and errors in our theories where more enthusiastic students see nothing but established truths. But I trust that I may have conveyed to many of my hearers a different impression. I have endeavoured to show that, even in what are usually considered the more trivial and superficial characters presented by natural objects, a whole field of new inquiry is opened up to us by the study of distribution and local conditions. And as regards man, I have endeavoured to fix your attention on a class of facts which indicate that the course of his development has been far less direct and simple than has hitherto been supposed ; and that, instead of resembling a single tide with its advancing and receding ripples, it must rather be compared to the progress from neap to spring tides, both the rise and the depression being comparatively greater as the waters of true civilisation slowly advance towards the highest level they can reach.

And if we are thus led to believe that our present knowledge of nature is somewhat less complete than we have been accustomed to consider it, this is only what we might expect ; for however great may have been the intellectual triumphs of the nineteenth century, we can hardly think so highly of its achievements as to imagine that, in somewhat less than twenty years, we have passed from complete ignorance to almost perfect knowledge on two such vast and complex subjects as the Origin of Species and the Antiquity of Man.

VIII.

THE DISTRIBUTION OF ANIMALS
AS INDICATING GEOGRAPHICAL CHANGES.[1]

Old Opinions on Continental Changes—Theory of Oceanic Islands—Present
and Past Distribution of Land and Sea—Zoological Regions—The
Palæarctic Region—The Ethiopian Region—The Oriental Region—Past
Changes of the Great Eastern Continent—Regions of the New World—
Past History of the American Continents—The Australian Region—
Summary and Conclusion.

THERE is a curious old book entitled *Restitution of De-
cayed Intelligence in Antiquities Concerning the Most
Noble and Renowned English Nation*, written in 1605,
by R. Verstegen. The fourth chapter treats " Of the
Isles of Albion, and how it is showed to have been continent or firm land with Gallia, now named France, since
the Flood of Noe ; " and after referring to several ancient
writers who had held this opinion but without giving any
reasons for it, the author proceeds to argue the point,
referring to the narrowness of the straits, their extreme
shallowness, the similarity of the opposite coasts both in
height and character, the meaning of the word " cliff "

---

[1] This is one of the Lectures on Scientific Geography delivered before the
Royal Geographical Society, but the introductory portion has been rewritten.
The original Lecture appeared in the *Proceedings* of the Society for September,
1877, under the title : " On the Comparative Antiquity of Continents, as
indicated by the Distribution of Living and Extinct Animals."

as being that which is cleft asunder, and other matters ; after which comes this quaint and interesting passage :—

"Another reason there is that this separation hath been made since the flood, which is also very considerable, and that is the patriarch Noe, having had with him in the Ark all sorts of beasts, these then, after the flood, being put forth of the ark to increase and multiply, did afterward in time disperse themselves over all parts of the continent or main land ; but long after it could not be before the ravenous wolf had made his kind nature known to man, and therefore no man unless he were mad, would ever transport of that race out of the continent into the isles, no more than men will ever carry foxes (though they be less damageable) out of our continent into the Isle of *Wight*. But our Isle, as is aforesaid, continuing since the flood fastened by nature unto the Great Continent, those wicked beasts did of themselves pass over. And if any should object that England hath no wolves on it they may be answered that Scotland, being therewith conjoined, hath very many, and so England itself sometime also had, until such time as King *Edgar* took order for the destroying of these throughout the whole realm."

The preservation of foxes for sporting purposes was evidently quite out of the range of thought at this not very distant epoch, and our author, in consequence, made a little mistake as to what men " ever " would do in the case of these noxious animals ; but his general argument is sound, and it becomes much strengthened when we take into consideration the smaller vermin, such as stoats, weasels, moles, hedgehogs, fieldmice, vipers, toads, and newts, which would certainly not *all* have been brought over by uncivilised man, even if any one of them might have been. But there is another reason why they were not so brought over. For on that supposition we should discover remains of fewer and fewer species as we go back into past times till at last when we reached the time of the first occupation of the country by man we should find none at all. But the actual facts are the very reverse of this. For the further we go back the more species of noxious and dangerous animals we discover, till in the time of the palæolithic (or oldest) prehistoric men, we find remains not only of almost every animal now living, but of many others still less likely to have been introduced by man's agency. Such are the mammoths, rhinoceroses, lions, horses, bears, gluttons, and many others; and it is equally impossible that these could all have swum across an arm of the sea, which although only about twenty miles wide in its narrowest past, is yet so influenced by strong tides and currents that it becomes as effective a barrier as many straits of double the width.

Owing, however, to the want of all definite ideas as to the mode by which the earth became stocked with animals and plants, the existence of identical species in countries separated by arms of the sea attracted very little attention till quite recent times. It is probable that Mr. Darwin was really the first person to see the full importance of the principle, for in his *Naturalist's Voyage Round the World*, he remarks, that "the South American character of the West Indian mammals seems to indicate that this archipelago was formerly united to the southern continent." Some years later, in 1845, Mr. George Windsor Earl called special attention to the subject by pointing out that the great Malay Archipelago may be divided in two portions, all the islands in the western half being united to each other and to the continent of Asia by a very shallow sea, and all having very similar productions, while many large animals, such as the elephant, rhinoceros, wild cattle, and tigers, range over most of them. We then come to a profoundly deep sea, and the islands of the eastern half of the archipelago are either surrounded by a deep sea or are connected by a shallow sea to Australia; and in this half the productions resemble those of Australia, marsupials being found in all the islands while the large quadrupeds of Asia are almost wholly unknown.

Theory of Oceanic Islands.—In 1859 the *Origin of Species* was published, and in the thirteenth chapter of this celebrated work Mr. Darwin put forth his views on oceanic islands or such as are situated far away from any continent and are surrounded by deep oceans. It had been up to this time believed that in most cases these islands were fragments of ancient continents; as an example of which we may refer to the Azores, Madeira, and the other Atlantic islands, which were thought to support the notion of an Atlantic or western extension of the European continent. In order to ascertain what was the condition of these islands when first discovered, Mr. Darwin searched through all the oldest voyages, and found that in none of them was a single native mammal known to exist, while in almost all of them frogs and toads were also absent. All the Atlantic isles from the Azores to St. Helena; Mauritius, Bourbon, and the other isles of the Indian Ocean; and the Pacific islands, east of the Fijis, as far as the Galapagos and Juan

Fernandez are thus deficient. They all of them, however, possess birds, and most of them bats; and whenever small mammalia, such as goats, pigs, rabbits, and mice have been introduced they have run wild and often increased enormously, proving that the only reason why such animals were not originally found there was the impossibility of them crossing the sea; while such as could fly over—birds, bats, and insects—existed in greater or less abundance. If, on the other hand, they had once formed part of the continent, it is impossible to believe that some of the smaller mammalia, as well as frogs, would not have continued to exist in the islands to the present day.

If we compare the productions of different islands, we meet with peculiarities which throw much light on the subject of distribution. In the Galapagos islands, between 500 and 600 miles from the west coast of South America, there are thirty-two species of land-birds, all but two or three being peculiar to the group. In Madeira, about 400 miles from the coast of Morocco, there are nearly twice as many land-birds as in the Galapagos, but only two of these are peculiar to the island, the rest being South European or N. African species. The Azores are 1,000 miles west of Portugal, and they contain twenty-two species of land-birds, every one of which is European except one bullfinch which is slightly different and forms a peculiar species. This remarkable difference in the proportion of peculiar species between the Galapagos and the Atlantic islands, is well explained by the theory that land-birds rarely fly directly out to sea, except when carried against their will by storms and gales of wind. Now the

Azores are situated in an especially stormy zone, and it is an observed fact that after every severe gale of wind some new bird or insect is seen on the islands. The Galapagos, on the contrary, are in a very calm sea where violent storms are almost unknown, and thus new birds from the mainland very rarely visit these islands. Madeira is less stormy than the Azores, but its comparative nearness makes up for this difference in the case of birds. In insects, however, the species of Madeira are much more peculiar (and more numerous) than those of the more distant Azores ; while those of the Galapagos are few, but all peculiar, and belonging to groups many of which are widely spread over the globe. All these facts are entirely in accordance with the view that oceanic islands have been peopled from the nearest continents by various accidental causes ; while they are entirely opposed to the theory that such islands are remnants of old continents and have preserved some portion of their inhabitants.

It is a curious fact, that land reptiles, such as snakes and lizards, are found in many islands where there are no mammalia or frogs ; and we therefore conclude that there must be some means by which their ova can be safely carried across great widths of sea. A single peculiar frog inhabits New Zealand, and some species are found in the Pacific islands as far eastward as the Fijis, but they are absent from all other oceanic islands. Snakes also extend to the Fijis, and there are two species in the Galapagos, but none in the other oceanic islands. Lizards, however, are found in Mauritius and Bourbon ; in New Zealand ; in all the Pacific islands, and in the Galapagos. It is clear then that next to Mammals, frogs and toads are most completely shut out by an ocean barrier ; then follow snakes, but as these are only found in the Galapagos and are very like South American species, they may possibly have been conveyed in boats or by floating trees. Lizards, however, are so wide-spread over almost all the warmer islands of the great oceans, that they must have some natural way of passing over, but the exact mode in which this is effected has not yet been discovered. Birds, as we have seen, are liable to be carried by winds and storms over great widths of sea, but this only applies to certain groups ; and large numbers which feed on the ground or which inhabit the depths of the forests, are almost as strictly confined to their respective countries by even a narrow arm of the sea as are the majority of the mammalia.

This sketch of the mode in which the various kinds of islands have been stocked with their animal inhabitants forms the best introduction to the study of those changes in our continents which have led to the existing distribution of animals. It demonstrates the importance of the sea as a *barrier* to the spread of all the higher animals ; and we are thus naturally led on to inquire, how far and to what extent such barriers have in past time existed between lands which are now united, and on the other hand what existing oceanic barriers are of comparatively recent origin. In pursuing this inquiry we shall have to take account of those grand views of the course of nature associated with the names of Lyell and Darwin—of the slow but never-ceasing changes in the physical conditions, the outlines and the mutual relations of the land-surfaces of the globe ; and of the equally slow and equally unceasing changes in the forms and structures of all organisms, to a great extent correlated with, and perhaps dependent on, the former set of changes. Combining these two great principles with other ascertained causes of distribution, we shall be enabled to deal adequately with the problem before us, and give a rational, though often only an approximative and conjectural, solution of the many strange anomalies we meet with in studying the distribution of living things.

Past and Present Distribution of Land and Sea.— Before proceeding to give details as to the distribution of animals, it is necessary to point out certain geographical features which have had great influence in bringing about the existing state of things.

The extreme inequality with which land and water is distributed has often been remarked, but what is less frequently noted is the singular way in which all the great masses of land are linked together. Notwithstanding the small proportion of land to water, the vast difference in the quantity of land in the northern and southern hemispheres, and the apparently hap-hazard manner in which it is spread over the globe, we yet find that no important area is completely isolated from the rest. We may even travel from the extreme north of Asia to the three great southern promontories—Cape Horn, the Cape of Good Hope, and Tasmania—without ever going out of sight of land ; and, if we examine a terrestrial globe, we find that the continents in their totality may be likened to a huge creeping plant, whose roots are at or around the North Pole, whose matted stems and branches cover a large part of the northern hemisphere, while it sends out in three directions great

offshoots towards the South Pole. This singular arrangement of the land surface into what is practically one
huge mass with diverging arms, offers great facilities for
the transmission of the varied forms of animal life over
the whole earth, and is no doubt one of the chief causes
of the essential unity of type which everywhere characterises the existing animal and vegetable productions of
the globe.

There is, moreover, good reason to believe that the
general features of this arrangement are of vast
antiquity ; and that throughout much of the Tertiary
period, at all events, the relative positions of our continents and oceans have remained the same, although
they have certainly undergone some changes in their
extent, and in the degree of their connection with each
other. This is proved by two kinds of evidence. In
the first place, it is now ascertained by actual measure-
ment that the depths of the great oceans are so vast
over wide areas, while the highest elevations of the land
are limited to comparatively narrow ridges, that the
mass of land (above the sea-level) is not more than $\frac{1}{36}$th
part of the mass of the ocean. Now we have reason to
believe that subsidence and elevation bear some kind of
proportion to each other, whence it follows that although
several mountain ranges have risen to great heights
during the Tertiary period, this amount of elevation
bears no proportion to the amount of subsidence required
to have changed any considerable area of what was
once land into such profound depths as those of the
Atlantic or Pacific Oceans. In the second place, we find
over a considerable area of all the great continents fresh-
water deposits containing the remains of land animals

and plants ; which deposits must have been formed in lakes or estuaries, and which therefore, speaking generally, imply the existence in their immediate vicinity of land areas comparable to those which still exist. The Miocene deposits of Central and Western Europe, of Greece, of India, and of China, as well as those of various parts of North America, strikingly prove this ; while the Eocene deposits of London and Paris, of Belgium, and of various parts of North and South America, though often marine, yet by their abundant remains of land-animals and plants, equally indicate the vicinity of extensive continents. For our purpose it is not necessary to go further back than this, but there is much evidence to show that throughout the Secondary, and even some portion of the Palæozoic periods, the land-areas coincided to a considerable extent with our existing continents. Professor Ramsay has shown[1] that not only the Wealden formation, and considerable portions of the Upper and Lower Oolite, but also much of the Trias, and the larger part of the Permian, Carboniferous, and Old Red Sandstone formations, were almost certainly deposited either in lakes, inland seas, or extensive estuaries. This would prove that, throughout the whole of the vast epochs extending back to the time of the Devonian formation, our present continents have been substantially in existence, subject, no doubt, to vast fluctuations by extension or contraction, and by various degrees of union or separation, but never so completely submerged as to be replaced by oceans comparable in depth with our Atlantic or Pacific.

[1] *Nature*, 1873, p. 333 ; *Quarterly Journal of the Geological Society*, 1871, pp. 189 and 241.

This general conclusion is of great importance in the study of the geographical distribution of animals, because it bids us avoid the too hasty assumption that the countless anomalies we meet with are to be explained by great changes in the distribution of land and sea, and leads us to rely more on the inherent powers of dispersal which all organisms possess, and on the union or disruption, extension or diminution, of existing lands —but always in such directions and to such a limited extent as not to involve the elevation of what are now the profoundest depths of the great oceans.

Zoological Regions.—We will now proceed to sketch out the zoological features of the six great biological regions; and will afterwards discuss their probable changes during the more recent geographical periods, in accordance with the principles here laid down.

The Palæarctic Region.

The Palæarctic, or North Temperate region of the Old World, is not only by far the most extensive of the zoological regions, but is the one which agrees least with our ordinary geographical divisions. It includes the whole of Europe, by far the largest part of Asia, and a considerable tract of North Africa; yet over the whole of this vast area there prevails a unity of the forms of animal life which renders any primary subdivision of it impossible, and even secondary divisions difficult. But besides being the largest of the great zoological regions, there are good reasons for believing this to represent the most ancient, and therefore the most important centre of the development of the higher forms of animal life,—and it is therefore well to consider it first in order.

In enumerating the most important animal groups characteristic of this and other regions, it must be clearly understood that such groups are not always absolutely confined to one region. Here and there they will often overlap the boundaries, while in other cases single species may have a wide distribution in one or more of the adjacent regions ; but this does not at all affect the main fact, that the group as a whole is very abundant and very widely spread over the region in question, while it is very rare, or confined to a very limited area in adjacent regions, and is therefore specially characteristic of the one as compared with other parts of the world. Bearing this in mind, we shall find that the Palæarctic region is well characterized by a considerable number of typical groups, although, as we shall presently see, it has in recent geological times lost much of its ancient richness and variety of animal life.

Among Mammalia the groups most characteristic of this region are the moles (Talpidæ), a family consisting of eight distinct genera which range over the whole region, but beyond it barely enter the Oriental region in North India, and the Nearctic region in North-West America ; camels, confined to the deserts of North Africa and Asia ; sheep and goats (Capra), only found beyond the region in the Nilgherries and Rocky Mountains ; several groups of antelopes, and many peculiar forms of deer ; hamsters (Cricetus), sand rats (Psammomys), mole rats (Spalax), and pikas (Lagomys), with several other forms of rodents. Wolves, foxes, and bears, are also very characteristic, though by no means confined to the region.

Among birds the most important group is certainly the small-sized, but highly-organized warblers (Sylviidæ), which, although almost universally distributed, are more numerous, and have more peculiar and characteristic genera here than in any other region. Most of our song-birds, and many of the commonest tenants of our fields, woods, and gardens, belong to this family; and identical or representative species are often found ranging from Spain to China, and from Ireland to Japan. The reedlings (Panuridæ), the tits (Paridæ), and the magpies (Pica), are also very characteristic; while among the finches (Fringillidæ), a considerable number of genera are peculiar. A large number of peculiar groups of grouse (Tetraonidæ), and pheasants (Phasianidæ) are also characteristic of this region. Although the reptiles and fresh-water fishes are comparatively few, yet many of them are peculiar. Thus, no less than 2 genera of snakes, 7 of lizards, and 16 of batrachia, are confined to the Palæarctic region, as well as 20 genera of fresh-water fishes.

The insects and land-shells offer their full proportion of peculiar types; but it would lead us beyond our special object to enter into details with regard to these less known groups of animals.[1]

[1] Details will be found in the Author's work on *The Geographical Distribution of Animals*.

The Ethiopian Region.

The Ethiopian region, consisting of Africa south of the Tropic of Cancer with Madagascar, is of very small area compared with the Palæarctic region ; yet owing to the absence of extreme climates, and the tropical luxuriance of a considerable portion of its surface, it supports a greater number and variety of large animals than any other part of the globe of equal extent. Much of the speciality of the region is, however, due to the rich and isolated fauna of Madagascar, the peculiarities of which may be set aside till we come to discuss the past history of the Ethiopian region.

Considering then, first, the zoological features of tropical and southern Africa alone, we find a number of very peculiar forms of mammalia.   Such are the golden moles, the Potamogale, and the elephant-shrews among Insectivora ; the hippopotami and the giraffes, among Ungulata ; the hyæna-like Proteles (Aard-wolf), and Lycaon (hyæna-dog), among Carnivora ; and the Aard-varks (Orycteropus) among Edentata.   These are all peculiar ; but among highly characteristic forms are the baboons, and several genera of monkeys and apes ; several peculiar Lemurs ; a great variety of the civet-family (Viverridæ), and of rodents ; peculiar genera of swine (Potamochærus and Phacochærus), and a greater abundance and variety of antelopes than are to be found in all the other regions combined.   But the Ethiopian region is strikingly distinguished from all others, not only by possessing many peculiar forms, but by the absence of a number of common and widely distributed groups of mammalia. Such are—the bears, which range over the whole northern hemisphere, and as far south as Sumatra in the eastern and Chili in the western hemisphere, yet they are totally wanting in Tropical and South Africa;—the deer, which are still more widely distributed, ranging all over North and South America, and over all Asia to Celebes and the Moluccas, yet they are totally absent from the Ethiopian region ; goats and sheep, true oxen (Bos), and true pigs (Sus), are also absent ; though as to the last there is some doubt, certain wild pigs having been observed, though rarely, in various parts of Tropical Africa, but it is not yet determined whether they are indigenous, or escaped from domestication. The absence of such wide-spread families as the bears and deer is, however, most important, and must be taken into account when we come to consider the geographical changes needed to explain the actual state of the Ethiopian fauna.

The birds are not proportionately so peculiar, yet there are many remarkable forms. Most important are the plantain-eaters, the ground-hornbills, the colies, and the anomalous secretary-bird ;—while among characteristic families there are numbers of peculiar genera of flycatchers, shrikes, crows, sun-birds, weaver-birds, starlings, larks, francolins, and the remarkable subfamily of the Guinea-fowls. There are not such striking deficiencies among birds as among mammals, yet there are some of importance. Thus, there are no wrens, creepers, or nut-hatches, and none of the widespread group comprising the true pheasants and jungle fowl—a deficiency almost comparable with that of the bears or the deer. Among the lower vertebrates there are 3 peculiar families of snakes and 1 of lizards, as well as 1 of toads and 3 of fresh-water fishes.

The Oriental Region.

The Oriental region comprises all tropical Asia east of the Indus, with the Malay Islands as far as Java, Borneo, and the Philippines. In its actual land-area it is the smallest region except the Australian; but if we take into account the wide extent of shallow sea connecting Indo-China with the Malay Islands, and which has, doubtless, at no distant epoch, formed an extension of the Asiatic Continent, it will not be much smaller than the Ethiopian region. Here we find all the conditions favourable to the development of a rich and varied fauna. The land is broken up into great peninsulas and extensive islands; lofty mountains and large rivers everywhere intersect it; while along its northern boundary stretches the highest mountain-range upon the globe. Much of this region lies within the equatorial belt, where the equability of temperature and abundance of moisture produce a tropical vegetation of unsurpassed luxuriance. We find here, as might be expected, that the variety and beauty of the birds and insects is somewhat greater than in the Ethiopian region; although, as regards mammalia, the latter is the most prolific, both in genera, species, and individuals.

The families of Mammalia actually peculiar to this region are few in number, and of limited extent. They are,—the Galeopithecidæ, or flying lemurs; the Tarsiidæ, consisting of the curious little tarsier, allied

to the lemurs ; and the Tupaiidæ, a remarkable group
of squirrel-like Insectivora.   There are, however, a considerable number of peculiar genera, forming highly
characteristic groups of animals—such as the various
apes, monkeys, and lemurs, almost all the genera of
which are peculiar ; a large number of civets and
weasels; the beautiful deer-like Chevrotains, often
called mouse-deer ; and a few peculiar antelopes and
rodents.   It must be remarked that we find here none
of those deficiencies of wide-spread families which were
so conspicuous a feature of the Ethiopian region—the
only one worth notice being the dormice (Myoxidæ), a
small family spread over the Palæarctic and Ethiopian
regions, but not found in the Oriental.

The birds of the Oriental region are exceedingly
numerous and varied, there being representatives of
about 350 genera of land-birds, of which nearly half
are peculiar.   Three families are confined to the region
—the hill-tits (Liotrichidæ), the green bulbuls (Phyl-
lornithidæ), and the gapers (Eurylæmidæ) ; while four
other families are more abundant here than elsewhere,
and are so widely distributed throughout the region as
to be especially characteristic of it.   These are—the
elegant pittas, or ground-thrushes (Pittidæ), the trogons
(Trogonidæ), the hornbills (Bucerotidæ), and the pheasants (Phasianidæ) ; represented by such magnificent
birds as the fire-backed pheasants, the ocellated pheasants, the Argus-pheasant, the pea-fowl, and the
jungle-fowl.

Reptiles are very abundant, but only 3 small families
of snakes are peculiar.   There are also 3 peculiar
families of fresh-water fishes.

Past Changes of the Great Eastern Continent.—
Having thus briefly sketched the main features of the
existing faunas of Europe, Asia, and Africa, it will be
well, while their resemblances and differences are fresh
in our memory, to consider what evidence we have of
the changes which may have resulted in their present
condition. All these countries are so intimately connected, that their past history is greatly elucidated by
the knowledge we possess of the tertiary fauna of
Europe and India ; and we shall find that when we
once obtain clear ideas of their mutual relations, we
shall be in a better position to study the history of
the remaining continents.

Let us therefore go back to the Miocene or middle
tertiary epoch, and see what was then the distribution
of the higher animals in these countries. Extensive
deposits, rich in animal remains of the Miocene age, occur
in France, Switzerland, Germany, Hungary, Greece; and
also in North-Western India at the Siwalik Hills, in
Central India in the Nerbudda Valley, in Burmah, and
in North China ; and over the whole of this immense
area we find a general agreement in the fossil mammalia,
indicating that this great continent was probably then,
as now, one continuous land. The next important geographical fact that meets us, is, that many of the largest
and most characteristic animals, now confined to the
tropics of the Oriental and Ethiopian regions, were then
abundant over much of the Palæarctic region. Elephants,
rhinoceroses, tapirs, horses, giraffes, antelopes, hyænas,
lions, as well as numerous apes and monkeys, ranged all
over Central Europe, and were often represented by a
greater variety of species than exist now. Antelopes

Y

were abundant in Greece, and several of these appear to have been the ancestors of those now living in Africa; while two species of giraffes also inhabited Greece and North-West India. Equally suggestive is the occurrence in Europe of such birds as trogons and jungle-fowl characteristic of tropical Asia, along with parrots and plaintain-eaters allied to forms now living in West Africa.

Let us now inquire what information Geology affords us of changes in land and sea at this period. From the prevalence of early tertiary deposits over the Sahara and over parts of Arabia, Persia, and Northern India, geologists are of opinion that a continuous sea or strait extended from the Bay of Bengal to the Atlantic Ocean, thus cutting off the Peninsula of India with Ceylon, as well as all tropical and South Africa from the great northern continent.[1] At the same time, and down to a comparatively recent period, it is almost certain that Northern Africa was united to Spain and to Italy, while Asia Minor was united to Greece, thus reducing the Mediterranean to the condition of two inland seas. We also know that the north-western Himalayas and some of the high lands of Central Asia were at such a moderate elevation as to enjoy a climate as mild as that which prevailed in Central Europe during the Miocene epoch,[2] and was therefore perhaps equally productive in animal and vegetable life.

---

[1] Mr. Searles V. Wood, "On the Form and Distribution of the Land-tracts during the Secondary and Tertiary Periods respectively," *Philosophical Magazine*, 1862.

[2] This part of the Himalayas was elevated during the Eocene period, and remains of a fossil *Rhinoceros* have been found at 16,000 feet elevation in Thibet.

We have, therefore, good evidence that the great Euro-Asiatic continent of Miocene times exhibited in its fauna a combination of all the main features which now characterise the Palæarctic, Oriental, and Ethiopian regions ; while tropical Africa, and such other tropical lands as were then, like the peninsula of India, detached and isolated from the continent, possessed a much more limited fauna, consisting for the most part of animals of a lower type, and which were more characteristic of Eocene or Secondary times. Many of these have no doubt become extinct, but they are probably represented by the remarkable and isolated lemurs of West Africa and Southern Asia, by the peculiar Insectivora of South Africa and Malaya, and by the Edentata of Africa and India. These are all low and ancient types, which were represented in Europe in the Eocene and early Miocene periods, at a time when the more highly specialised horses, giraffes, antelopes, deer, buffaloes, hippopotami, elephants, and anthropoid apes had not come into existence. And if these large herbivorous animals were all wanting in tropical Africa in Miocene times, we may be quite sure that the large felines and other carnivora which prey upon them were absent also. Lions, leopards, and hyænas can only exist where antelopes, deer, or some similar creatures abound ; while smaller forms allied to the weasels and civets would be adapted to a country where small rodents or defenceless Edentata were the chief vegetable-feeding mammalia.

If this view is correct (and it is supported by a considerable amount of evidence which it is not possible here to adduce), all the great mammalia which now seem so specially characteristic of Africa—the lions, leopards, and hyænas,—the zebras, giraffes, buffaloes, and antelopes,—the elephants, rhinoceroses, and hippopotami,—and perhaps even the numerous monkeys, baboons, and anthropoid apes,—are every one of them comparatively recent immigrants, who took possession of the country as soon as an elevation of the old Eocene and Miocene sea-bed afforded a passage from the southern borders of the Palæarctic region. This event probably occurred about the middle of the Miocene period, and it must have effected a vast change in the fauna of Africa. A number of the smaller and more defenceless of the ancient inhabitants must have been soon exterminated, as surely as our introduced pigs, dogs, and goats, exterminate so many of the inhabitants of oceanic islands ; while the new comers finding a country of immense extent, with a tropical climate, and not too much encumbered with forest vegetation, spread rapidly over it, and thenceforth, greatly multiplying, became more or less modified in accordance with the new conditions. We shall find that this theory not only accounts for the chief specialities, but also explains many of the remarkable deficiencies of the Ethiopian fauna. Thus, bears and deer are absent, because they are comparatively late developments, and were either unknown or rare in Europe till late Miocene or Pliocene times ; while, on the other hand, the immense area of open tropical country in Africa has favoured the preservation of numerous types of large mammalia which have perished in the deteriorated climate and diminished area of Europe.

Our knowledge of the geology of Africa is not sufficiently detailed to enable us to determine its earlier history with any approach to accuracy. It is clear, however, that Madagascar was once united with the southern portion of the Continent, but it is no less clear that its separation took place before the great irruption of large animals just described ; for all these are wanting, while lemurs, insectivora, and civets abound,— the same low types which were once the only inhabitants of the mainland. It is worthy of note, that south temperate Africa still exhibits a remarkable assemblage of peculiar forms of mammalia, birds, and insects,—the two former groups mostly of a low grade of organisation ; and these, taken in connection with the wonderfully rich and highly specialised flora of the Cape of Good Hope, point to the former existence of an extensive south-temperate land in which so many peculiar types could have been developed. Whether this land was separated or not from Equatorial Africa, or formed with it one great southern continent, there is no sufficient evidence to determine.

Turning now to tropical Asia, we find a somewhat analogous series of events, but on a smaller scale and with less strongly-marked results. At the time when tropical and South Africa were so completely cut off from the great northern continent, the peninsula of India with Ceylon was also isolated ; and it seems probable that their union with the continent took place at a somewhat later period. The ancient fauna of this south-Asiatic island may be represented by the slow Loris, a peculiar type of lemurs, some peculiar rats (Muridæ), and perhaps by the Edentate scaly ant-eater ; by its Uropeltidæ, a peculiar family of snakes, and by many peculiar genera of snakes and lizards, and a few peculiar amphibia. On the other hand, we must look upon the monkeys, the large carnivora, the deer, the antelopes, the wild pigs, and the elephants, as having overrun the country from the north; and their entrance must, no doubt, have led to the extermination of many of the lower types.

But there is another remarkable series of changes which have undoubtedly taken place in Eastern Asia in Tertiary times. There is such a close affinity between the animals of the Sunda Islands and those of the Malay Peninsula and Siam; and between those of Japan and of Northern Asia, that there can be little doubt that these islands once formed a southern and eastern extension of the Asiatic continent. The Philippines and Celebes perhaps also formed a part of this continent; but if so, the peculiarity and poverty of their mammalian fauna shows that they must have been separated at a much earlier period.[1] The other islands probably remained united to the continent till the Pliocene period. The result is seen in the similarity of the flora of Japan to that which prevailed in Europe in Miocene times; while in the larger Malay Islands we find, along with a rich flora developed under long-continued equatorial conditions of uniform heat and moisture, a remnant of the fauna which accompanied it, of which the Malay tapir, the anthropoid apes, the tupaias, the galeopitheci or flying lemurs, and the sun-bears, may be representatives.

There is another very curious set of relations worthy of our notice, because they imply some former com-

---

[1] For a full account of the evidence and conclusions as to these islands see the author's *Geographical Distribution of Animals*, vol. i. pp. 345, 359, 426, 436.

munication between the Malay Islands, on the one hand, and South India with Ceylon, on the other. We find, for example, such typical Malay forms as the Tupaia, some Malay genera of cuckoos and Timaliidæ, some Malayan snakes and amphibia. The remarkable genus Hestia among butterflies, and no less than seven genera of beetles of purely Malay type,[1] all occurring either in Ceylon only or in the adjacent parts of the Peninsula, but in no other part of India. These cases are so numerous and so important, that they compel us to assume some special geographical change to account for them. But directly between Ceylon and Malaya there intervenes an ocean-depth of more than 15,000 feet; and besides the improbability of so great a subsidence, of which we have no direct evidence, a land communication of this kind would almost certainly have left more general proofs of its existence in the faunas of the two countries. But, when in Miocene times a sub-tropical climate extended into Central Europe, it seems probable that the equatorial belt of vegetation accompanied by its peculiar fauna, would have been wider than at present, extending perhaps as far as Burma. If then the shallow northern part of the Bay of Bengal had been temporarily elevated during the late Miocene or Pliocene epochs, a few Malayan types may have migrated to the Peninsula of India; and have been preserved only in Ceylon and the Nilgherries, where the climate still retains somewhat of its equatorial character and the struggle for existence is somewhat less severe than in the northern part of the region, which is so much more productive in varied forms of life.

[1] For details see *Geographical Distribution of Animals*, vol. i. p. 327.

There are also indications hardly less clear, of some communication between India and Malaya on the one hand, and Madagascar on the other ; but as these indications depend chiefly on resemblances in the birds and insects, they do not imply that any land connection has occurred. If, as seems probable, the Laccadive and Maldive Islands are the remains of a large island or indicate a western extension of India, while the Seychelles, with the shallow banks to the south-east and the Chagos group are the remains of other extensive lands in the Indian Ocean, we should have a sufficient approximation of these outlying portions of the two continents to allow a certain amount of interchange of such winged groups as birds and insects, while preventing any intermixture of the mammalia.

The presence of some African types (and even some African species) of mammals in Hindostan appears to be due to more recent changes, and may perhaps be explained by a temporary elevation of the comparatively shallow borders of the Arabian Sea, admitting of a land passage from North-East Africa to Western India.

There remains to be considered the supposed indications of a very ancient communication between Africa, Madagascar, Ceylon, Malaya, and Celebes, furnished by the occurrence over this extensive area of isolated forms of the Lemur tribe. The anomalous range of this group of animals has been thought to require for its explanation the existence of an ancient southern continent which has been called Lemuria, but a consideration of all the facts does not seem to warrant such a theory. Had such a continent ever existed we are sure that it must have disappeared long before the Miocene period, or it would

assuredly have left more numerous and widespread
indications of the former connections of these distant
lands than actually exist. And when we go back to
the Eocene period we are met by the interesting discovery of an undoubtedly Lemurine animal in France,
and what are supposed to be allied forms in North
America. This proof of the great antiquity and wide
range of lemurs is quite in accordance with their low
grade of development; while the extreme isolation and
specialization of many of the existing types (of which
the Aye-aye of Madagascar is a wonderful example),
and their scattered distribution over a wide tropical
area, all suggest the idea that these are but the remnants of a once extensive and widely distributed group
of animals, which, in competition with higher forms,
have preserved themselves either by their solitary and
nocturnal habits, or by restriction to ancient islands,
like Madagascar, where the struggle for existence has
been less severe. Lemuria, therefore, may be discarded
as one of those temporary hypotheses which are useful
for drawing attention to a group of anomalous facts,
but which fuller knowledge shows to be unnecessary.

Regions of the New World.—We will now pass across
the Atlantic to the Western Hemisphere, and consider
first the Nearctic region, or temperate North America,
whose present and past zoological relations with the rest
of the world are of exceeding interest.

If we omit such animals as the musk-sheep (Ovibos),
which is purely Arctic, and the peccaries (Dicotyles),
which are hardly less distinctly tropical, the land-

mammalia of North America are not very numerous; and they can be for the most part divided into two groups, the one allied to the Palæarctic, and the other to the Neotropical fauna. The bears, the wolves, the cats, the bison, sheep and antelope, the hares, the marmots, and the pikas, resemble Palæarctic forms; while the racoons, skunks, opossum, and vesper-mice are now more peculiarly Neotropical. There are also many genera which are altogether peculiar and characteristic of the region, as the prong-horn antelope (Antilocapra), the jumping-mouse (Jaculus), five genera of pouched rats (Saccomyidæ), the prairie dogs (Cynomys), the tree porcupines (Erethizon), and some others.

Birds present the same mixture of the two types; but the wild turkeys (Meleagris), the passenger pigeon (Ectopistes), the crested quails (Lophortyx, &c.), the ruffed grouse (Cupidonia), and some other groups of less importance, are peculiar; while the family of the wood warblers (Mniotiltidæ) is so largely developed that it may claim to be more characteristic of North than of South America.

Reptiles and Amphibia present a number of peculiar types; while no less than five peculiar families of fresh-water fishes would alone serve to mark out this as distinct from every other part of the world.

Considering the evident affinity between the Nearctic and Palæarctic regions, there are here some curious deficiencies of groups which are common and widely-spread in the latter. Thus hedgehogs, wild horses and asses, swine, true oxen, goats, dormice, and true mice are absent; while sheep and antelopes are only represented by solitary species in the Rocky Mountains.

Among birds, too, we have such striking deficiencies as the extensive families of flycatchers, starlings, and pheasants.

Turning now to the Neotropical region, comprising all South America and the tropical parts of the northern continent, we find that the Old World types have still further diminished, while a number of new and altogether peculiar forms have taken their place. Insectivora have wholly disappeared with the exception of one anomalous form in the greater Antilles; bears are represented by one Chilian species; swine are replaced by peccaries; the great Bovine family are entirely unknown; the camel tribe are confined to the Southern Andes and the south temperate plains; deer are not numerous; and all the varied Ungulata of the Old World are represented only by a few species of tapirs. These great gaps are, however, to some extent filled up by a variety of interesting and peculiar types. Two families of monkeys (Cebidæ and Hapalidæ) differ in many points of structure from all the Quadrumana of the eastern hemisphere. There is a peculiar family of bats—the vampyres; many peculiar weasels and Procyonidæ; a host of peculiar rodents, comprising five distinct families, among which are the largest living forms of the order; and a great number of Edentata, comprising the families of the sloths, armadillos, and ant-eaters; and lastly, a considerable number of the marsupial family of opossums. As compared with the Old World, we find here a great abundance and variety of the lower types, with a corresponding scarcity of such higher forms as characterise the tropics of Africa and Asia.

In birds we meet with corresponding phenomena. The most abundant and characteristic families of the Old World tropics are replaced here by a series of families of a lower grade of organisation, among which are such remarkable groups as the chatterers (Cotingidæ), the manakins (Pipridæ), the ant-thrushes (Formicariidæ), the toucans (Rhamphastidæ), the motmots (Momotidæ), and the humming-birds (Trochilidæ), the last perhaps the most remarkable and beautiful of all developments of the bird-type. Parrots are numerous, but these, too, are mostly of peculiar families; while pheasants and grouse are replaced by curassows and tinamous, and there are an unusual number of remarkable and isolated forms of waders.

Reptiles, amphibia, fresh-water fishes, insects, and land-shells, are all equally peculiar and abundant; so that South America presents, on the whole, an assemblage of curious and beautiful natural objects, unsurpassed—perhaps even unequalled—in any other part of the globe.

Past History of the American Continents.—We will now proceed to examine what is known of the past history of the two American continents, and endeavour to determine what have been their former relations to each other and to the Old World, and how their existing zoological and geographical features have been brought about. And first let us see what knowledge we possess of the past relations of North America with the Eastern continents.

If we go back to that recent period termed the Post-Pliocene—corresponding nearly to the Post-Glacial period and to that of pre-historic man in Europe—we find at once a nearer approximation than now exists between the Nearctic and Palæarctic faunas. North America then possessed several large cats, six distinct species of the horse family, a camel, two bisons, and four species of elephants and mastodons. A little earlier, in the Pliocene period (although fossil remains of this age are scanty), we have in addition the genus Rhinoceros, several distinct camels, some new forms of ruminants, and an Old-World form of porcupine. Further back, in the Miocene period, we find a Lemuroid animal, numerous insectivora, a host of carnivora, chiefly feline and canine, a variety of equine and tapirine forms, rhinoceroses, camels, deer, and an extensive extinct family—the Oreodontidæ—allied to deer, camels, and swine. There are, however, no elephants. In the still earlier Eocene period most of the animals were peculiar, and unlike anything now living, but some were identical with European types of the same age, as Lophotherium and the family Anchitheridæ.

These facts compel us to believe that at distinct epochs during the Tertiary period the interchange of large mammalia between North America and the Old World has been far more easy than it is now. In the Post-Pliocene period, for example, the horses, elephants, and camels of North America and Europe were so closely allied that their common ancestors must have passed from one continent to the other, just as we feel assured that the common ancestors of the American and European bison, elk, and beaver, must have so migrated. We have further evidence in the curious fact that certain groups appear to come into existence in the one continent much later than in the other. Thus cats, deer, mastodons, true horses, porcupines, and beavers, existed in Europe long before they appeared in America; and as the theory of evolution does not admit the independent development of the same group in two disconnected regions to be possible, we are forced to conclude that these animals have migrated from one continent to the other. Camels, and perhaps ancestral horses, on the other hand, were more abundant and more ancient in America, and may have migrated thence into Northern Asia.

There are two probable routes for such migrations. From Norway to Greenland by way of Iceland and across Baffin Bay to Arctic America, there is everywhere a comparatively shallow sea, and it is not improbable that during the Miocene period, or subsequently, a land communication may have existed here. On the other side of the continent, at Behring Straits, the probability is greater. For here we have a considerable extent of far shallower sea, which a very slight elevation would convert into a broad isthmus connecting North America and North-East Asia. It is true that elephants, horses, deer, and camels would, under existing climatal conditions, hardly range as far north as Greenland and Alaska; but we must remember that most mysterious yet indisputable fact of the luxuriant vegetation, including even magnolias and other large-leaved evergreens, which flourished in these latitudes during the Miocene period; so that we have all the conditions of favourable climate and abundant food, which would render such interchange of the animals of the two continents not only possible, but inevitable, whenever a land communication was effected; and there is reason to believe that this favourable condition of things continued in a diminished degree during a portion of the succeeding Pliocene period.

We must not forget, however, that the faunas of the two continents were always to a great extent distinct and contrasted—such important Old-World groups as the civets, hyænas, giraffes, and hippopotami, never passing to America, while the extinct Oreodontidæ, Brontotheridæ, and many others are equally unknown in the Old World. This renders it probable that the communication even in the north was never of long continuance ; while it wholly negatives the theory of an Atlantis bridging over the Atlantic Ocean in the Temperate Zone at any time during the whole Tertiary period.

But the past history of the North-American fauna is complicated by another set of migrations from South America, which, like those from the Old World, appear to have occurred at distant intervals, and to have continued for limited periods. In the Post-Pliocene epoch, along with elephants and horses from Europe or Asia we find a host of huge sloths and other Edentata, as well as llamas, capybaras, tapirs, and peccaries, all characteristic of South America. Some of these were identical with living species, while others are closely allied to those found fossil in Brazilian caves and other deposits of about the same age, while nothing like them inhabited the Old World at the same period. We are therefore quite sure that they came from some part of the Neotropical region ; but the singular fact is, that in the preceding Pliocene epoch none of them are found in North America. We conclude, therefore, that their migration took place at the end of the Pliocene or beginning of the Post-Pliocene epoch, owing to some specially favourable conditions, but that they rapidly disappeared, having left no survivors. We must, however, study the past history of South America in order to ascertain how far it has been isolated from or connected with the northern continent.

Abundant remains of the Post-Pliocene epoch from Brazilian caves show us that the fauna of South America which immediately preceded that now existing had the same general characteristics, but was much richer in large mammalia and probably in many other forms of life. Edentata formed the most prominent feature ; but instead of the existing sloths, armadillos, and ant-eaters, there were an immense variety of these animals, some of living genera, others altogether different, and many of them of enormous size. There were armadillos as large as the rhinoceros, while the megatherium and several other genera of extinct sloths were of elephantine bulk. The peculiar families of South American rodents —cavies, spiny-rats, and chinchillas—were represented by other species and genera, some of large size ; and the same may be said of the monkeys, bats, and carnivora. Among Ungulata, however, we find, in addition to the living tapirs, llamas, peccaries, and deer, several species of horse and antelope, as well as a mastodon, all three forms due probably to recent immigration from the northern continent.

Further south, in Bolivia, the Pampas, and Patagonia, we also find abundant fossil remains, probably a little older than the cave fauna of Brazil, and usually referred to the newer part of the Pliocene period. The same families of rodents and Edentata are here abundant, many of the genera being the same but several new ones also appearing. There are also horses, peccaries, a mastodon, llamas, and deer ; but besides these there are a number of altogether peculiar forms, such as the Macrauchenia, allied to the Tapir and Palæotherium ; the Homalodontotherium, allied to the miocene Hyracodon of North America ; and the Toxodontidæ, a group of very large animals having affinities to Ungulates, rodents, Edentata, and Sirenia, and therefore probably the representative of a very ancient type.

Here then we meet with a mixture of highly developed and recent, with low and ancient types, but the latter largely predominate ; and the most probable explanation seems to be that the same concurrence of favourable conditions which allowed the megatherium and megalonyx to enter North America also led to an immigration of horses, deer, mastodons, and many of the Felidæ into South America. These inter-migrations appear to have taken place at several remote intervals, the northern and southern continents being for the most part quite separated, and each developing its own peculiar forms of life. This view is supported by the curious fact of a large number of the marine fishes of the two sides of Central America being absolutely identical—implying a recent union of the two oceans and separation of the continents —while the mollusca of the Pacific coast of America bear so close a relation to those of the Caribbean Sea and the Atlantic coasts, as to indicate a somewhat more remote but longer continued sea-passage. The straits connecting the two oceans were probably situated in Nicaragua and to the south of Panama, leaving the highlands of Mexico and Guatemala united to North America.

Around the Gulf of Mexico and the Caribbean Sea there is a wide belt of rather shallow water, and during the alternate elevations and subsidences to which this region has been subjected, the newly raised land would afford a route for the passage of immigrants between North and South America. The great depression of the ocean, believed to have occurred during the Glacial period (caused by the locking-up of the water in the two polar masses of ice), may perhaps have afforded the opportunity for those latest immigrations which gave so striking a character to the North American fauna in Post-Pliocene times.

Among the changes which South America itself has undergone, perhaps the most important has been its separation into a group of large islands. Such a change is clearly indicated by the immense area and low elevation of the great alluvial plains of the Orinoko, Amazon, and La Plata, as well as by certain features in the distribution of the existing Neotropical fauna. A subsidence of less than 2,000 feet would convert the highlands of Guiana and Brazil into islands separated by a shallow strait from the chain of the Andes. When this occurred the balance of the land was probably restored by an elevation of the extensive submerged banks on the east coast of South America, which in South Brazil and Patagonia are several hundred miles wide, embracing the Falkland Islands, and reaching far to the south of Cape Horn.

Looking, then, at the whole of the evidence at our command, we seem justified in concluding that the past histories of North and South America have been different, and in some respects strongly contrasted. North America was evidently in very early times so far connected with Europe and Asia as to interchange with those continents the higher types of animal life as they were successively developed in either hemisphere. These more perfectly organised beings rapidly gained the ascendency, and led to the extinction of most of the lower forms which had preceded them. The Nearctic has thus run a course parallel to that of the Palæarctic region, although its fauna is, and perhaps always has been, less diversified and more subject to incursions of lower types from adjacent lands in the southern hemisphere.

South America, on the other hand, has had a history in many respects parallel to that of Africa. Both have long existed either as continents or groups of large islands in the southern hemisphere, and for the most part completely separated from the northern continents ; and each accordingly developed its peculiar types from those ancestral and lowly-organised forms which first entered it. South America, however, seems to have had a larger area and more favourable conditions, and it remained almost completely isolated till a later period. It was therefore able to develop a more-varied and extensive fauna of its own peculiar types, and its union with the northern continent has been so recent, and is even now maintained by so narrow an isthmus, that it has never been overrun with the more perfect mammalia to anything like the extent that has occurred in Africa. South America, therefore, almost as completely as Australia, has preserved for us examples of a number of low and early types of mammalian life, which, had not the

z 2

entire country been isolated from the northern continent during middle and late Tertiary times, would long since have become extinct.

The Australian Region.—There only remains for us now to consider the relation of the island-continent of Australia to Asia and South America, with both which countries it has a certain amount of zoological connection.

Australia, including New Guinea (which has in recent times been united with it), differs from all the other continents by the extreme uniformity and lowly organisation of its mammalia, which almost all belong to one of the lowest orders—the marsupials. Monkeys, carnivora, insectivora, and the great and almost ubiquitous class of hoofed-animals, are all alike wanting; the only mammals besides marsupials being a few species of a still lower type—the monotremes, and a few of the very smallest forms of rodents—the mice. The marsupials, however, are very numerous and varied, constituting 6 families and 33 genera, of which there are about 120 known species. None of these families is represented in any other continent; and this fact alone is sufficient to prove that Australia must have remained almost or quite isolated during the whole of the Tertiary period.

In birds there is, as we might expect, less complete isolation; yet there are a number of very peculiar types. About 15 families are confined to the Australian region, among which are the paradise-birds, the honey-suckers, the lyre-birds, the brush-tongued lories, the mound-makers, and the cassowaries.

Our knowledge of the former mammalian inhabitants

of Australia is imperfect, as all yet discovered are from Post-Tertiary or very late Tertiary desposits. It is interesting to find, however, that all belong to the marsupial type, although several are quite unlike any living animals, and some are of enormous size, almost rivalling the mastodons and megatheriums of the northern continents. In the earliest Tertiary formation of Europe remains of marsupials have been found, but they all belong to the opossum type, which is unknown in Australia; and this supports the view that no communication existed between the Palæarctic and Australian regions even at this early period. Much farther back, however, in the Oolite and Trias formations, remains of a number of small mammalia have been found which are almost certainly marsupial, and bear a very close resemblance to the Myrmecobius, a small and very rare mammal still living in Australia. An animal of somewhat similar type has been discovered in rocks of the same age in North America; and we have, therefore, every reason to believe, that it was at or near this remote epoch when Australia, or some land which has been since in connection with it, received a stock of mammalian immigrants from the great northern continent; since which time it has almost certainly remained completely isolated.

The occurrence of the marsupial opossums in America has been thought by some writers to imply an early connection between that continent and Australia; but the fact that opossums existed in Europe in Eocene and Miocene times, and that no trace of them has been found in North or South America before the Post-Pliocene period, renders it almost certain that they entered America from Europe or North Asia in middle or late Tertiary times, and have flourished there in consequence of a less severe competition with highly-developed forms of life.

The birds of Australia and South America only exhibit a few cases of very remote and general affinity, which are best explained by the preservation in each country of once wide-spread types, but is quite inconsistent with the theory of a direct union between the two countries during Tertiary times.

Reptiles are even more destitute of proofs of any such connection than even mammalia or birds; but in amphibia, fresh-water fishes, and insects the case is different, all these classes furnishing examples of the same families or genera inhabiting the temperate parts of both continents. But the fact that such cases are confined to these three groups and to plants, is the strongest possible proof that they are not due to land-connection; for all these organisms may be transmitted across the ocean in various ways. Violent storms of wind, floating ice, drift-wood, and aquatic birds, are all known to be effective means for the distribution of these animals or their ova, and the seeds of plants. All of them too, it must be noted, are to a considerable degree patient of cold; the reverse being the case with true reptiles and land-birds, which are essentially heat-loving; so that the whole body of facts seems to point rather to an extension of the Antarctic lands and islands reducing the width of open sea, than to any former union, or even close approximation of the Australian and South American continents.

Summary and Conclusion.

Let us now briefly review the conclusions at which we have arrived. If we look back to remote Tertiary times, we shall probably find that all our great continents and oceans were then in existence, and even bore a general resemblance to the forms and outlines now so familiar to us. But in many details, and especially in their amount of communication with each other, we should observe important changes. The first thing we should notice would be a more complete separation of the northern and the southern continents. Now, there is only one completely detached southern land—Australia ; but at that period Africa and South America were also vast islands or archipelagos, completely separated from their sister continents. Examining them more closely, we should observe that the great Euro-Asiatic continent had a considerable extension to the south-east, over what are now the shallow seas of Japan, China, and Java. In the south-west it would include Northern Africa, the Mediterranean then forming two inland seas ; while to the west and north-west it would include the British Isles, and perhaps extend even to Iceland and Greenland. As a balance to these extensions, much of Northern Siberia and North-Western Asia may have been under water ; the peninsula of India would be an island with a con- siderable south-west extension over what are now the Laccadive and Maldive coral-reefs. The Himalayas would be a moderate range of hills ; the great desert plateau of Central Asia a fertile plain ; the greater part of the continent would enjoy a tropical or sub-tropical climate, while even the extreme north would support a luxuriant vegetation. This great continent would abound in animal life, and would be especially remarkable for its mammalia, which would comprise ancestral forms of all our existing higher types, along with a number of those lower grades of organisation (such as lemurs and opossums) now found chiefly in the southern hemisphere.

Connected with this continent by what is now Behring Straits and the Sea of Kamschatka, we should find North America, perhaps somewhat diminished in the east, but more extensive in the south and north, and abounding as now with great inland lakes which were situated to the west of the present lake district. This continent seems to have had a less tropical climate and vegetation than prevailed in the eastern hemisphere, but it supported an almost equally varied though very distinct fauna. Ancestral horses no larger than dogs; huge tapir-like and pig-like animals; strange forms allied to rhinoceroses; the Dinocerata—huge horned animals allied to elephants and to generalised Ungulata; and the Tillodontia, still more unlike anything now living, since they combined characters now found separated in the carnivora, the Ungulata, and the rodents. Ancestral Primates, allied to both the lemurs and the South American monkeys, also inhabited this continent.

The great land masses of the northern hemisphere thus appear to have possessed between them all the higher types of animal life; and these seem to have been developed for a time in one continent and then to have been in part transferred by migration to the other, where alone they have sometimes maintained themselves.

Thus, the elephants and the camels appear to have descended from what were once exclusively American types, while the opossums were as certainly European. Many groups, however, never passed out of the continent in which they originated—the civets, hyænas, and the giraffes being wholly eastern, while the Oreodontidæ and Brontotheridæ were no less exclusively western.

South America seems to have been united to the northern continent once at least in Secondary or early Tertiary times, since it was inhabited in the Eocene period by many forms of mammalia, such as rodents, felines, and some ancient forms of Ungulata. It must also have possessed the ancestors of the Edentata (though they have not yet been discovered), or we should not find such a variety of strange and gigantic forms of this order in later Tertiary deposits in this part of the world only. During the greater part of the Tertiary period, therefore, South America must have been separated from the North and protected from incursions of the higher forms of mammalia which were there so abundant. Thus only does it seem possible to understand the unchecked development of so many large but comparatively helpless animals as the Edentata of the Pampas and the Brazilian caves—a development only comparable with that of the Australian marsupials, still more completely shut off from all competition with higher forms of life.

In Africa the evidence of a long period of insulation is somewhat more complex and less easily apparent, but, it seems to me, equally conclusive. We have first, the remarkable fauna of Madagascar, in which lemurs and insectivora predominate, with a few low forms of carnivora ; but none of the higher animals, such as apes, antelopes, buffaloes, rhinoceroses, elephants, lions, leopards, and hyænas, which swarm on the continent. The separation of Madagascar from Africa must therefore have occurred before these important groups existed there. Now, we know that all these large animals lived in Europe and Asia during late Miocene times, while lemurs are only known there during the Eocene period, and were probably more abundant in late Mesozoic times. It is almost certain, therefore, that Southern Africa must have been cut off from Europe and Asia during the whole intervening period, or the same development of high forms and extinction of low would have gone on in the one country as in the other. The persistence of a number of low and isolated types in South and West Africa, which are probably a remnant of the ancient fauna of the country, is also favourable to this view. At the time we are considering, therefore, we look upon tropical and South Africa, with Madagascar, as forming a completely isolated land or archipelago ; while the Seychelles and Chagos banks, with Bourbon and Mauritius, perhaps, formed another island or group permanently separated from the larger masses. The extra-tropical portion of South Africa was also probably more extensive, affording an area in which its remarkable flora was being developed.

Turning to Australia, we should probably find it, at this remote period, more extensive than it is now, including in its area New Guinea and some of the adjacent islands, as well as Tasmania ; while another extensive land probably occupied the site of the New Zealand group. It may be considered certain that, whatever elevations and subsidences these countries may have undergone, they have not been connected either with Asia, Africa, or South America during the whole Tertiary period.

In conclusion, I would especially remark that the various changes in the outlines and mutual relations of our continents, which I have now endeavoured to establish, must not be supposed to have been all strictly contemporaneous. Some may have been a little earlier or a little later than others ; some changes may have been slower, others more rapid ; some may have had but a short duration, while others may have persisted through considerable geological periods. But, notwithstanding this uncertainty as to details, the great features of the geographical revolutions which I have indicated, appear to be established by a mass of concurring evidence ; and the lesson they teach us is, that although almost the whole of what is now dry land has undoubtedly once lain deep beneath the waters of the ocean, yet such changes on a great scale are excessively slow and gradual; so that, when compared with the highest estimates of the antiquity of the human race, or even with that of most of the higher animals, our existing continents and oceans may be looked upon as permanent features of the earth's surface.

ERRATUM.

AT page 59 I have said that there are only three or four species of Mimosa which are sensitive. This is a mistake, as the greater portion of the species in the extensive genus Mimosa, as well as some species of several other genera of Leguminosæ, and also of Oxalidaceæ, possess this curious property. I cannot find, however, that any one has suggested in what way the sensitiveness may have been useful to the species which first acquired it. My guess at an explanation may therefore induce botanists who are acquainted with the various species in a state of nature, to suggest some better solution of the problem.

INDEX.

A.

Abrus precatoria, perhaps a case of mimicry, 226
Absorption-colours or pigments, 183
Acræidæ, warning colours of, 174
Adaptive characters, 150, 155
Affinities, how to determine doubtful, 148
African large mammalia, recent immigrants, 323
Allen, Mr. Grant, on protective colours of fruits, 225
Alpine flowers, why so beautiful, 232
Amboyna, large sized butterflies of, 258
American monkeys, 118
American Continents, past history of, 332
Ancient races of North and South America, 298
Andaman Islands, pale butterflies of, 260
    white-marked birds of, 263
Anderson, Mr. W. Marshall, on cranium from N. American mound, 296
Andes, very rich in humming-birds, 139
Animal colours, how produced, 184
    life in tropical forests, 70
Anthribidæ, 95
Ants, wasps, and bees, 80
    numbers of, in India and Malaya, 81—88
    destructive to insect-specimens, 85
    and vegetation, special relation between, 89
Apatura and Heterochroa, resemblance of species of, 257
Apes, 116
Aqueous vapour of atmosphere, its influence on temperature, 9
    quantity at Batavia and Clifton, 10
Arctic plants, large leaves of, 236
    flowers and fruits brightly coloured, 237
Areca palm, 45
Arenga saccharifera, 43

Argus-pheasant, wonderful plumage of, 205
Arums, 48
Assai of the Amazon, 43
Auckland Isles, handsome flowers of, 238
Audubon, on the ruby humming-birds, 130, 137
Australian Region, mammalia of, 340
    birds of, 340
    extinct fauna of, 341
    its supposed union with S. America, 341
Azara, on food of humming-birds, 135

B.

BAMBOOS, 52
    uses of, 53—58
Bananas, wild, 47
Banana, 48
Barber, Mrs. on colour changes of pupa of *Papilio nireus*, 168
Barbets, 105
Bark, varieties of in tropical forests, 33
Barometer, range of, at Batavia, 24
Batavia, Meteorology of, 4
    and London, diagram of mean temperatures, 5
    greatest rainfall at, 24
    range of barometer at, 24
Bates, Mr. on climate at the Equator, 24
    on scarcity of forest-flowers on Amazon, 61
    on animal life in Amazon valley, 70
    on abundance of butterflies at Ega, 75
    on importance of study of butterflies, 78
    on leaf-cutting ants, 86
    on blind ants, 88
    on bird-catching spider, 97
    on use of toucan's bill, 106
    on large serpents, 115
    on the habits of humming birds, 132